ERROR TRAPS

ELMAR LUTTER

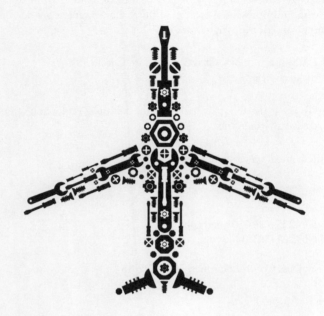

ERROR TRAPS

HOW HIGH-PERFORMING TEAMS

LEARN TO AVOID MISTAKES

IN AIRCRAFT MAINTENANCE

Advantage | Books

Published by Advantage Books, Charleston, South Carolina.
An imprint of Advantage Media.

ADVANTAGE is a registered trademark, and the Advantage colophon is a trademark of Advantage Media Group, Inc.

Printed in the United States of America.

10 9 8 7 6 5 4 3 2 1

ISBN: 978-1-64225-784-7 (Paperback)
ISBN: 978-1-64225-783-0 (eBook)

Library of Congress Control Number: 2023913248

Book design by Megan Elger.

This publication is designed to provide accurate and authoritative information in regard to the subject matter covered. It is sold with the understanding that the publisher is not engaged in rendering legal, accounting, or other professional services. If legal advice or other expert assistance is required, the services of a competent professional person should be sought.

The views and opinions expressed in this book belong solely to the author and do not necessarily reflect the views of LHT or any other company referenced within. The author has attempted to be as accurate as possible, but please forgive any mistakes related to technical details.

Some manufacturers, such as Boeing and Airbus, have been used as examples of error traps in the book. No offense or fingerpointing was intended by sharing these episodes. Though no efforts will be enough to fully eliminate all error traps, the manufacturers' and the whole community's commitment to make aviation safe and affordable remains unquestionable.

Advantage Books is an imprint of Advantage Media Group. Advantage Media helps busy entrepreneurs, CEOs, and leaders write and publish a book to grow their business and become the authority in their field. Advantage authors comprise an exclusive community of industry professionals, idea-makers, and thought leaders. For more information go to **advantagemedia.com**.

To Washington SyCip (1921–2017)

CONTENTS

FOREWORD

AIR TRAFFIC SAFETY HAS IMPROVED continuously in recent decades. The airplane is the safest means of transport today. In theory, you could fly every day for 11,000 years before a fatal accident would occur.

This level of safety is the result of joint efforts by aircraft, engine and component manufacturers, the certification authorities, and the training and education programs of the airlines and maintenance providers. The cooperation and the exchange of experience of all those involved—including between competitors—to increase system reliability and thus also security has led to this excellent level of safety over the past decades.

It is unique that a single industry can come together to achieve such outstanding results through continuous improvements, which benefit everyone involved. The motivation for this willingness to cooperate becomes understandable as soon as one considers the consequences of a fatal accident with its dramatic effects. Everyone involved is aware that there must be no mistakes that would lead to an accident. The consequences for the affected airline and those responsible at all levels, but also for the entire industry, are unforeseeable.

On the one hand, the level of safety achieved is reassuring, but on the other hand, there is a risk of believing oneself too much in a safety net. The danger that this can lead to carelessness and thus to errors, and that potential hazards are generously overlooked, is latent. The aim of this book is to make people aware of this and to maintain a clear focus and "alert level" when working on the aircraft.

Errors in the aircraft systems and when working on the aircraft can occur anytime and anywhere. The well-known "Murphy's Law" says: "Every mistake that is possible will also occur." It is extremely important to keep this fact in mind at all times. It means, for example, that even an aircraft that has just been delivered by the manufacturer or a completely overhauled aircraft from the D-check[1] can have errors. A reasonable distance and impartiality should therefore always be maintained when working on the aircraft and resolving issues, whether the aircraft is new or old.

You always have to reckon with errors in every phase of an aircraft's life. In my professional practice, I have seen time and again that, on the one hand, unexpected problems arose, especially with new aircraft, and, on the other hand, system problems suddenly surfaced with very mature aircraft types after more than twenty years in operation, which very quickly affected the entire fleet.

Therefore, there is no alternative but to focus, respect, be responsible, and remain impartial when solving problems on the aircraft.

I would like to add one more factor to Murphy's Law: "Any error that has once occurred will occur again." This is a very important statement that deserves special attention.

Immediate learning from recognized errors is important to prevent repeat cases. This applies to detected errors in the aircraft systems as well as to work errors in operation.

---

1   A D-check is the biggest hangar visit in an aircraft's life, typically lasting several weeks and due every 12 years, see also Glossary.

Any mistake that is repeated is a missed opportunity to avoid a possible delay, an aircraft on ground (AOG), or the same work error. Therefore, rapid communication and clarification with all colleagues who were not involved in solving the problem is extremely important in order to avoid a repeat of a complaint about the aircraft or the work process.

Accidents with fatal consequences have already happened in the past in which the failure of an important component that triggered or contributed to the accident had already occurred as a problem with one or more airlines before and was therefore known. Reluctance to act and downplaying problems, combined with other factors, can lead to serious risks and failure of the entire safety system.

Aviation and thus flying are safe, but when working on the aircraft you should always be aware of your responsibility and actively accept it.

I am very pleased that Elmar Lutter has dealt with this question intensively in his book and with a practical perspective. Experience and routine are necessary and desirable for efficient work on the aircraft. At the same time, experience and routine must not lead to overconfidence and superficiality. Maintaining focus at work and respect for the task are nowhere more important than in aviation.

I wish the book will have many attentive readers who will heed the recommendations.

AUGUST WILHELM HENNINGSEN
Former Chief Executive Officer, Lufthansa Technik AG (2001–2015)

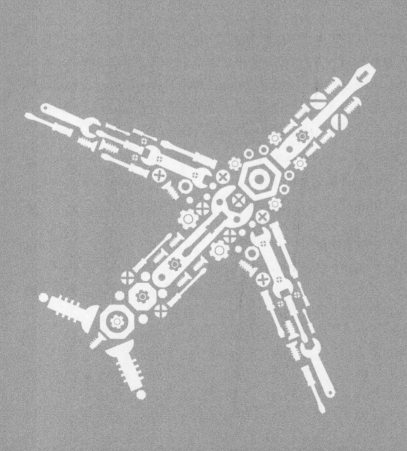

INTRODUCTION

Hard work is irrelevant without integrity.

—ROLANDO PAULINO,

SECTION MANAGER, TECHNICAL SERVICES

WHEN SOMETHING GOES WRONG in aircraft maintenance, the mechanics involved typically go through a miserable time. They get tested for drugs and are possibly suspended from work. They feel guilty that they have let their team down. In extreme cases, they consider resigning or even taking their lives.

And then you look at *how* things go wrong, and it is often the familiar pattern: People underestimate how easily mistakes can happen in maintenance and how seemingly small mistakes can result in big damage to aircraft and people—in short, how exposed they are every day. And you replace a mechanic with a pilot, planner, or manager and maintenance with any other field, and it could be very similar.

I am writing this with the misery I have seen in mind. As their leader, you suffer with them. Have I done enough to see it coming and mitigate the risk? Obviously not. The situations in which things go wrong seem to have much in common—to an extent which makes the accident or incident appear even more predictable and preventable. I started to look at these situations as "error traps." My mission became to find ways to reduce the risk the people in my organization, including myself, are exposed to. The risk of a predictable mistake, which is not on their mind.

In an ideal world, we make things intuitive to use and foolproof, such as the famous "Norman doors," which let you know whether to push or pull.[1] There are many examples in aviation where things function intuitively, like the little wheel on the landing gear lever, a guarded switch not to be pushed accidentally, or a built-in self-test of the aircraft systems. However, these features are far from foolproof and there are arguably more examples of *less* intuitive design. Ideally, systems or processes are designed in a "fail-safe" way to withstand an erroneous input or at least behave gracefully when parts of them malfunction. There are some good examples, yes, but in general, the remaining exposure is immense.

Starting with the closing of the V2500 fan cowl doors (FCDs), which can easily go wrong, I collected examples of error traps for an internal training program with the same name. In these trainings, I saw that people are surprisingly unaware of the patterns in which things tend to fail. I felt it is my obligation to do more to prepare my colleagues for the possibility that they end up in a life-changing error, which looks totally predictable to the rest of the world—an *error trap*.

In aviation, an error can cost you your job, or, in extreme cases, your life, and not to mention the lives of others. An error can cost you your job, even if you have not *caused* damage. Now, *what causes what* and who or what is to blame is a controversial topic in safety management. Let's look at an example.

In October 2018, Lion Air Flight 610 crashed and killed all on board—the first fatal accident of a Boeing 737MAX. The investi-

---

1   Don Norman, *The Design of Everyday Things: Revised and Expanded Edition*, 2nd ed. (Hachette: Basic Books, 2013).

gation report lists twenty-five recommendations without assigning blame or root causes (let alone *the one* cause).[2]

The immediate cause, which set the chain of events in motion and distinguished it from a normal flight on a normal day, was the miscalibration of a component (the AOA sensor) in a repair shop in Florida, a type of fault that possibly happens several times per day somewhere in the system without bringing down an airliner. Even within that repair shop, it wasn't a simple linear chain of events initiated or concluded by a classical human error, but a rather systemic issue to operate test equipment without detailed instruction of use.

No single individual acted outside their envelope of normal performance variation, let alone recklessly. It was an accident waiting to happen. Still, one of the involved actors in the cockpit or on the ground could have managed to avert the catastrophe on the fateful two days after the faulty component was installed. The pilots on the day before the accident especially could have taken into consideration the erratic behavior of the airplane more seriously than making a routine logbook entry about fault messages. That is, of course, speculation, and even more so would be the question of what would have happened then instead. Nobody would say, though, that these pilots have caused the crash. On the contrary, what happened at Lion Air was not remarkable and different from what happens everywhere in the worldwide aviation system without any consequences most of the time.

It is sometimes, arguably always, difficult to say what *caused* an accident or incident.

It is now widely accepted that Boeing and the FAA have a prominent role in the effort to preclude such a disaster from happening

---

2   KNKT, "Aircraft Accident Investigation Report - PT. Lion Mentari Airlines - Boeing 737-8 (MAX); PK-LQP - 29 October 2018" (Jakarta, October 2019).

again. Many decisions at Boeing, since the 1990s, have contributed to the design issues which have resulted in an almost inevitable accident.[3] This is the field of safety science: the many factors detached in space and time which cause complex socio-technical systems to fail. If safety science is successful, the whole system will see less accidents.

But there is also another perspective: The pilots, especially those on the penultimate flight, and maintenance could have saved this flight. This is not to say they have *caused* the accident. Saved *this* flight, but maybe not at the same time another or the second B737MAX crash, Ethiopian Airlines Flight 302, some months later. This is the field of error control. If successful, our own small sphere of influence may see less accidents but not the system as a whole, at least not directly.

Safety science and error control have different ways to look at accidents and their causes, especially when it comes to preserving and further extending the achievements of safety in aviation. The common ground here seems to be the need for capable *leadership* at all levels.[4] I think it is fair to say that we don't blame the pilots or maintenance technicians for Boeing's shortcomings in the example above, but at the same time, that we accept those shortcomings as a fact of life, at least on our next shift and, thus, expect certain behaviors from pilots and maintenance technicians which take these realities into account.

Our angle here is error control. What can we do in our sphere of influence to avoid mishaps, which is important for two reasons: First,

---

3   Peter Robison, *Flying Blind: The 737 MAX Tragedy and the Fall of Boeing* (New York: Penguin Business, 2022).

4   For examples of different angles on safety leadership, see: Robert J. de Boer, *Safety Leadership: A Different, Doable and Directed Approach to Operational Improvements* (Boca Raton, FL: CRC Press, 2021), https://doi.org/10.1201/9781003143338; Karl E. Weick and Kathleen M. Sutcliffe, *Managing the Unexpected: Sustained Performance in a Complex World* (San Francisco, CA: John Wiley & Sons, 2015); Tony Kern, *Blue Threat: Why to Err Is Inhuman* (Pygmy Books, 2009).

the chain of events ending in damages can be intercepted by certain behaviors flowing from specific attitudes. Second, and this is the very mission of this book, avoiding predictable errors will reduce exposure to become collateral damage in mishaps initiated elsewhere, either as a direct victim of design flaws or other suboptimal decisions, or indirectly like Lion Air's maintenance. The technicians who released Flight 610 to service were suspended immediately after the accident. It is unknown whether they have been reinstated, but the investigation report also revealed their shortcomings in troubleshooting, whether these errors could be seen as contributing factors or not. I call this *exposure*. Many people are just not aware of how exposed they are, every day.

This book is about error control. But it's not so much about techniques and best practices, as the title might suggest. It's more about this one modest insight that aircraft maintenance (and life in general) is a hostile environment, in which accidents will always look like they were waiting to happen. Which tools you use then to ensure you have done your very best to keep every aircraft you touch as safe as possible for your team on the ground and the pilots and passengers in the air, and how you choose to put in place your defenses, is entirely up to you.

It is my hope that this discussion, based on years of collected knowledge and experience, can in some measure, improve airline safety and reduce the personal liability that every person in the aviation system is exposed to every day.

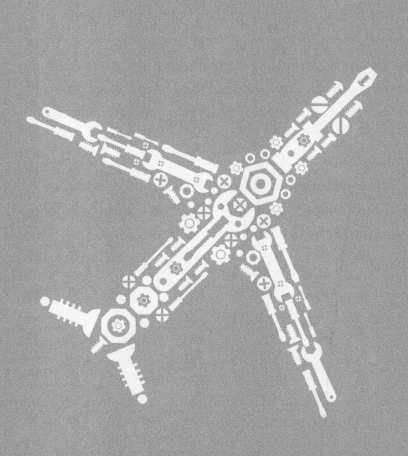

CHAPTER 1

The Mother of All Error Traps

V2500 FAN COWL DOORS

"Why are people having problems?" they wonder. "You are being too logical," I say. "You are designing for people the way you would like them to be, not for the way they really are."

—THE DESIGN OF EVERYDAY THINGS BY DON NORMAN

ROMANIA, NOVEMBER 2011. Bucharest Băneasa International Airport (BBU) is the home base for ten Airbus A320 airplanes of the fast-growing Romanian fleet of a pan-European low-cost carrier. Our team consists of twenty-five Hungarian line maintenance technicians, delegated from our home base in Budapest. Tough guys. Super experienced. It's minus 10, minus 20 degrees Celsius on this cold, dark Romanian night, and they simply turn up the collars of their jackets and shrug it off.

Tonight is the weekly A-check for one of the ten planes in the fleet. The A-check is a more substantial work package requiring all hands on deck, so that the routine night-stop work can continue in parallel. At least tonight's A-check can be done in a hangar hosted by a local aerospace company. It's unheated, so full outdoor gear remains on, but at least it serves as a buttress from the howling winds and blowing snow. The crew works under pressure and mostly alone on their tasks all night, as there is no spare capacity on an A-check. Not having the plane ready for its first flight in the early morning is not an option, and everyone knows they must stay on task to meet the deadline.

Just before 4:00 a.m., all that is left is the final close-up and paperwork before the plane is towed out from the hangar. One of the crew, I'll call him Gábor, worked on the V2500 engines. He checked the integrated drive generator (IDG) oil level and differential pressure indicator among other tasks requiring him to open FCDs. The FCDs are like the hood of a car; they separate the engine from the elements but theoretically, the plane could fly without them. To open and close them is as simple as closing the hood of the car, with one difference: The FCDs must be secured by four latches underneath the engine. This requires the mechanic to lie on their back beneath the engine and work upward.

Gábor performs the task, locking the FCDs and the plane is ready for tow out.

At 6:00 a.m., the aircraft takes off as scheduled for its first flight of the day. As the plane ascends, screams erupt from the passengers on the left side of the plane as they watch the wind peel back the large FCD and rip it off its hinges. It flies past their windows, smashing against the wing and the tailplane before it soars out of sight. The engine is now fully exposed and begins to billow what looks like smoke. It's not long before images of the terror appear on social media. A PR nightmare for this growing low-cost carrier in a promising new market.

The plane lands safely back at BBU, but what a disaster. Two million in repairs and $8 million in lost revenues during the six weeks the plane will be grounded for repairs. Besides the FCDs and the damage to the fuselage and leading edges of the wing and the horizontal stabilizer, the biggest headache is the pylon cantilever where the FCDs are attached to the plane and which are badly bent upward.

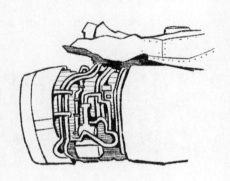

So, did Gábor fully lock the FCDs? That's how he remembers it, but whether he just lowered the doors or how many latches he closed couldn't be recovered in the investigation. It wouldn't be unusual to close the access panels only as the very last step before signing off the whole work package which often requires "pre-close" inspections by the final release person or

V2500 fan cowl doors (FCDs) bent upwards at the pylon cantilever and disintegrated. The FCDs were not closed properly, so they couldn't withstand the airflow during take-off.

a power-up self-test. Maybe Gábor just loosely engaged one hook to hold them together until he would get the final signal from his chief. Or he may have felt the time pressure and didn't notice that one latch didn't fully engage, and the spring-loaded mechanism wasn't fully over-centered. That would have been enough.

The slightest gap between the doors and the adjacent cowlings allows air to flow under the doors and rips them off at takeoff. One latch not perfectly closed is enough for that. But at least Airbus is efficient in providing parts, tools, and repair assistance. To our surprise, this was not the first time it had happened. Before our case, similar accidents had occurred.

In each instance, the plane landed safely, so safety was not considered comprised. While the FCDs' lightweight honeycomb structures can dent the aircraft's structure behind them, they cannot destroy it. But how can we be sure? That leads us to Takeaway No. 1: Even an event as terrifying and expensive as the loss of the FCDs during takeoff is not bad enough for an immediate redesign, let alone grounding of the fleet, because flight safety had not been severely impacted. Or, in short:

TAKEAWAY NO. 1: If a design problem doesn't bring down a plane, chances are that it won't get fixed.

In fairness, there were attempts to fix the problem. The latches were painted fluorescent orange and were modified with a small weight to make the latches reliably hang down when not engaged. If

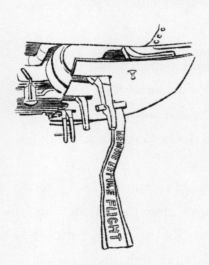

"Remove Before Flight" streamer (or "tethered flag"). Part of the key to open the foremost V2500 fan cowl door latch. The flag hangs down and can be seen when standing in front of the engine.

you stand in front of the engine, though, you cannot see the latches hanging down, unless you step back a few meters or kneel. In addition, a procedure was introduced to make a tech log entry when you open the FCDs, so that it shouldn't be possible to forget to close them. The pilot wouldn't take the aircraft with an open entry but to no avail.

Between 1992 and 2018, we've counted twenty-six cases of FCD losses of V2500 engines. Twenty-six! And then, finally, in 2018, the comprehensive fix arrived (or so we thought). The latches were redesigned to require a key to open and when the latch is open it locks the key, so that the key with its red "Remove before flight" flag remains there until the latch is closed again. This flag hangs further down than the latch and can be seen easily when in front of the engine, e.g., during a pre-flight walk-around. One key for each side is stored in the cockpit, with the idea that the mechanic may forget the tech log entry but not the key required to open the doors.

I call this an *error trap*. The design of this part doesn't reliably preclude a simple error with dramatic consequences, in the cold, dark Romanian night or not. Knowing how often the FCDs must be opened in standardized maintenance programs, we can estimate that one incident per year in the worldwide fleet with V2500 engines would come with 40,000 successful operations. You can look at one in 40,000 opportunities as very seldom or very often. We'll review that perspective later.

The FCD design is not the only error trap that exists. It is an example of an error trap induced by design or a design weakness if you will. We find these not only in the design of aircraft and their parts, but also in the design of tools and procedures. And if we look closer, we can also find error traps in the way we make decisions, and that how we arrive at bad decisions is not completely random. And if we consider all the bad decisions we make in life, isn't it that we can see patterns? If a friend tells you about some calamity of their own making, isn't it that you have seen that before? It appears to me that life is full of error traps. But let's get back to the error traps inherent in aircraft design.

DESIGN FLAWS: CAN WE OVERCOME THEM?

Takeaway No. 1 is that designs can be weak, and they don't get fixed, at least not immediately, unless something catastrophic occurs, like in the case of the B737MAX crashes in 2018 and 2019. In the case of the B737MAX, the aircraft design based on the 1960s B737 model was stretched to accommodate the big present-day engines. Unfortunately, that made the airplane harder to fly and required additional means to stabilize it, which turned out to confuse the pilots. The "fix" or defense for the additional stabilization needed was a flight-control software called Maneuvering Characteristics Augmentation System (MCAS), which turned out to be a key cause of the crashes that killed 346 people. Why is it so hard to get designs correct the first time? That leads us to:

TAKEAWAY NO. 2: Design costs are consistently underestimated.

Intuitively, we think designs of products should be finished before delivery, tested in all circumstances, and robust in the hands of consumers. You expect your laptop charger to work with any voltage your hotel room outlet can provide. For aircraft design, this has never been the case. The market is too small, too competitive, and the initial qualification, out of sight for the consumer, is too expensive. For similar reasons, it takes longer to go back to the moon than it took to go there in the first attempt. Much longer.

A new aircraft design like the A350 or B787 costs in the range of $15 billion. A huge price tag which does, however, not include the type of testing and finetuning that you are accustomed to with your iPhones or your toaster. Newer aircraft are much better designed and built, to be sure. Just look at the uncontained engine failure which forced a Qantas A380 to return to Singapore in 2010. Its structural and system damages would in earlier times have taken down the aircraft. But the expectations and complexities have risen faster than technology can keep up with. The consequence is that we cling to 1960s designs, like the B737MAX, or attempt new design fixes for every part, every system on the aircraft. These frequent design fixes after entry into service are called "Service Bulletins." Not all of them are implemented by the airline, unless they are mandatory, as often retrofit costs outweigh quantifiable benefits. Modifications to enhance maintainability are even harder to justify.

And so, we must live with unfinished and weak designs. Twenty-six FCD losses! At least for a new aircraft type design, this scenario triggered a reconsideration and the A321neo with PW1100 engines now features an electrical switch on the door latch, which warns the flight deck of an unclosed FCD.

No key is needed anymore. We call these countermeasures *defenses*, and if they are effective, the error trap ceases to exist. The early defenses on the V2500 FCD were, for one reason or another, not effective, causing recurring incidents. With the introduction of the new electrical switches on the PW1100 FCD latches as an effective defense, it seems that the improper or forgotten closing of

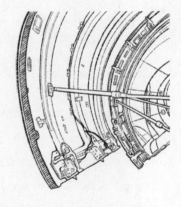

New FCD design, used with A321neo PW1100. The latches have now electrical switches, which signal an open position to the flight deck.

the FCD latches has been fixed and is no longer an error trap: time will tell. In general:

TAKEAWAY NO. 3: Tasks which easily go wrong lack effective "defenses" which results in error traps.

Our profession is full of error traps, because it is a small market with extraordinarily high development costs. But this is just the tip of the iceberg. Other fields are not so much better, and life, as such, has error traps for us at every turn. So, what can we do? Aiming to avoid all error traps sounds like shooting for the moon. However, doing nothing cannot be an option. If a grave error isn't bad enough, a *predictable* grave error would clearly add insult to injury, and the last thing we want is to be viewed as incompetent professionals in front of our peers, right?

So, to battle error traps, let's consider:

TOOL NO. 1: Competence: Master the basics, read and understand the guidance.

Competence means mastering the vocational basic skills, so that you can perform your job reliably under normal circumstances. If you are not there (yet), then the discussion of error traps may not make complete sense to you (yet). Remember, one in 40,000 is a long way to go. But it is also not that black and white. Aircraft maintenance can be quite forgiving in the sense that small errors, most of the time, don't matter much. The next touch will happen long before any dangerous state could manifest. If you miss an inspection, the next one will catch it. If you use some wrong fluid or compound, adverse impact would be very slow, if any. Most mechanics have never received any bad feedback about their work. Either a mistake was not detected before correction or the next station did not bother to report it. That's reality. If a tool is left behind, which can be traced back, that's embarrassing. Or, if after a heavy-maintenance visit something obvious is missing or malfunctioning, that's embarrassing. But that's the exception. Often, small maintenance errors go undetected, and people may accept them as good enough, but a mindset of "good enough" cannot be justified.

We must learn from these small errors and deviations so that bad habits do not become part of the team norms. One wrong part may bring down an airliner, as we have seen with the accidents of a Lockheed L-1011 in 1983 and a BAC-111 in 1990, in which fatalities were narrowly avoided. Wrong parts were installed in these cases, because the mechanics didn't check for the correct part number as per standard practice. They used parts that looked similar to those which they replaced in the BAC-111 windshield replacement case, or they used parts which previously came in a different configuration in the L-1011 magnetic chip detectors (MCD) replacement case. Aiming for precision in all manual tasks provides you the skillset and mindset to avoid error traps.

Case 1: Lockheed L-1011. An in-flight turn-around was caused by all three engines failing on a flight from the USA to the Caribbean when the oil leaked out of each. The oil leak was caused by missing O-rings on the MCD. They were missing because the mechanic had not noticed that the new chip detectors were not fitted with O-rings in the usual way. All work was performed outside in darkness, where a black O-ring was difficult to see. Until that night, chip detectors had always come with O-rings attached, even though the mechanic had to sign for both components. The new packaging still said they were ready for use.

Case 2: BAC-111. During industrial action at the airline, a maintenance manager changed a windshield himself. He had not performed this task for two years but checked the Maintenance Manual and it looked straightforward. He replaced eighty of the eighty-four bolts. The correct bolts were A211-8D, although A211-7D were on the old windshield. He matched the old bolts to new ones in a stores bin, but chose A211-8C, which was the correct length but the wrong thread. They engaged in the holes, but he used the wrong torque in setting them. Also because of the awkward posture required he could not see the bolts tighten. On the first flight, the windshield blew out, severely injuring the pilot and forcing an in-flight turn-around.[5]

What is more immediate, though, is reading and *understanding* the guidance. This is really the first hurdle and the first line of defense. Manuals are only as good as the designs, which means they are not perfect. Sometimes they are hard to understand or even outright wrong. But more often, they are just not given the attention required. The tool at hand here is to *not* attempt a task that you don't understand at first reading, especially when it is written in your second language. You must make the effort and time to review and, if necessary, stop and ask to make sure you fully understand.

For me, understanding the manuals is a bit like the elephant in the room. Too often, I have seen people go ahead with a shallow grasp of what is described, sometimes cherry-picking for the bits they can't go without, like torque values or precision measurement equipment (PME) readings. Instead, we must aim for precision which requires

---

5 Colin G. Drury, *Human Factors in Aircraft Maintenance,* State University of New York at Buffalo, accessed July 6, 2023, https://skybrary.aero/sites/default/files/bookshelf/2504.pdf.

finding time to look up or investigate what we haven't understood at first. Many details can be safely "deferred" to later for a full under-standing, but it requires a little extra effort once you have the time. For example, the opening of the FCDs requires you to use a key "with a *tethered* flag." This English word may have never crossed your way before, but you can also safely ignore it for now and look it up in a quiet moment. The ability to intuit which words are essential to know at the moment and which are not will come with time. Your reading skills, especially in a second language, are becoming a success factor, in particular your reading speed and your ability to switch between a normal, reflective mode and a high-tempo mode.

To really understand what we are doing or what we are supposed to be doing (the guidance) is our first line of defense. From my experi-ence, people are too often working with a 98 percent understanding, and a misunderstanding of whether to go left or right from one point; whether an alternate procedure is allowed or not in this case; whether a tool is specifically required or can be a general-purpose one; can, at times, make all the difference.

In our case, the closing of the V2500 FCDs, fully understanding the manual, helps avoid its error trap. Today (not yet in 2011, as the new latches were only introduced in 2018), the manual states: "DO NOT TRY TO REMOVE THE KEY FROM THE LATCH WHEN THE LATCH IS OPEN." During a spot-check in 2019, I discovered that is what exactly had happened. The key was removed on the first side to use it again on the other. On another aircraft (with both keys correctly installed), I asked a mechanic whether he knew what the key was for. He guessed it was for precluding unauthorized access. That is, of course, not the intent. But he also doesn't have to know if he just follows the instructions provided in the manual. Which he did. I also asked him whether he knew how to remove the key, and he said, "of course, that's easy with a screwdriver." Which he didn't. That's good, but leaves us with a slight concern that this defense is perhaps weaker than we thought.

TOOL NO. 2: Awareness: Be in the moment: know what is coming.

When *competence* is our first line of defense, *awareness* or alertness is our second. We know that sometimes situations are not ideal, and errors are more likely. We have seen those patterns before, and when we can see the potential for errors, we shift to high alert. A well-functioning early-warning system can go a long way. If we are fully in the moment and we keep our focus firmly on the task, chances are that we can adapt to unforeseen circumstances or developments successfully.

Two years of pandemic and virtual meetings have taught us to pay attention to our mic being muted or not, especially, when we have something important to say. It still happens, "you are muted," but we have all improved, because we've learned to pay extra attention to it before we speak. The same holds true the moment before pressing the send button on an email and discovering that we are about to hit reply all rather than our intended forward. Some of us have learned that the hard way. The FCD example required that same attention and focus. One of the mechanics back in Romania in 2011 told me that closing the latches correctly is so important and delicate that he always taps them after the move, a small bump with the bottom of the fist, to detect whether it is firmly closed.

This type of "mindfulness" has become a popular concept for improving results by being "fully in the moment" and shutting out all distractions. Preparation for the unexpected is a proven error-control technique. In psychological research studies, participants have been better able to solve tasks by being aware of what types of challenges may come their way.[6]

As outlined in *Managing Maintenance Error: A Practical Guide* by James Reason and Alan Hobbs, one study compared the performance of two experimental groups engaged in solving equations on the computer. Both groups were subjected to intermittent distracting images at the top of their screen, but each group was told a different method with which to deal with those distractions. Group one was instructed to develop their own mental technique that would enable them to ignore the images (disruption inhibition). Group two was instructed to tell themselves that they would do the task at hand as best they could despite the distractions (task facilitation).

Which group fared better? Group one scored significantly higher than group two and both groups scored better than the control groups who did nothing to prepare for the distractions. Focusing only on the task at hand and staying on the lookout for challenges that arise is clearly effective, but it is also difficult to maintain 24/7. Improving our ability to consciously switch between high alert and autopilot will in turn improve our performance and ability to adapt to an evolving situation.

Vigilant *awareness* becomes our second line of defense. You understand the task and you perform it competently, and you also know which step counts most or what could go wrong. For the FCDs, the manual reminds you, "Do not leave this job after just closing the fan cowls, continue on to secure the latches. If you are called away prior to latching, then either re-open one cowl door or latch the

---

6   James Reason and Alan Hobbs, *Managing Maintenance Error: A Practical Guide* (London: CRC Press, 2017).

latches before walking away from this engine." The FCD should not be left closed while unlatched under any circumstances.

Being fully competent and fully in the moment is great. But can you do it every single minute, in every single scenario? Can anybody be that sure? That brings us to our third line of defense: *Compliance.*

TOOL NO. 3: Compliance: Respect the defenses: read and *follow* the guidance.

Many view aircraft maintenance as an overregulated and over-specified industry. The opening of the FCD has sixteen pages of instruction and ten warnings for a one-minute task. With all the safety precautions and defenses, the task may become ten minutes. That's the world we work in. We sometimes spend more time on assuring safety—preparing for the job, putting all the precautionary measures in place, closing everything safely after the job, documenting it, and hopefully improving it for next time—than on the work itself.

That is a fact we must accept, because those measures come from previous accidents and incidents. Sometimes it annoys us. Sometimes we feel it is over the top. But at other times, when our competence and our mindfulness must carry us over unknown territory (there is more we don't know than we do know), the stored experience of over a hundred years of aviation will help us. We must respect the risks which those additional steps (defenses) in the procedure are trying to mitigate. We must *not* ignore, circumvent, or forget them even when we are rushed. The guidance is there for us, we just have to follow it.

It takes effort to remind ourselves to simply follow the existing guidance. Several years ago, I was so often at Frankfurt Airport (FRA) that I always knew the best way between any two points. Nowadays, I am just following the signs, the guidance, which is there, and I use my brain for more important tasks than airport navigation. FRA insiders will object, however, that blindly following the signage between terminals may have you unnecessarily leaving the security area. And they have a point. The same thing can happen when you blindly follow your car's satnav system. We obviously cannot and should not switch off our brains completely. Even when we, by default, follow the guidance, a small piece of our mental software should monitor for common sense violations, like leaving the main road for a much smaller one without a cue, or known error traps, like heading from Terminal A to Terminal B at FRA.

And so, compliance is a hot topic. We are often proud of "our way." We see small deviations as marks of our creativity and superiority. The strict uniform standards are for the foot soldiers, we are the special forces. All of us rebel at times. And what is more, we may look like hypocrites, as really nobody can and would comply with every little detail to the letter or nothing would get done. Have you read and understood *everything* before you sign off to have read and understood everything? So, who are we to ask people for full compliance? Let's postpone that discussion for a while and agree to take compliance—read and *follow* the guidance—as a tool to avoid making the same mistakes others made. We know that we cannot be perfectly competent and alert and mindful every single moment: to fill that gap, we rely on *compliance*.

TOOL NO. 4: Teamwork: Look after one another.

Teamwork is meant here not in the sense that certain jobs need more than one person, like to carry a slide up the aircraft stairs or to use a headset person on the ground, but rather teamwork as a tool to protect us from error traps. Watching out for each other, offering to help our colleagues before they make a grave mistake, is *teamwork*. And it is our last line of defense.

Often it is easier to see shortcomings in others than in ourselves. We are humans after all. To avoid an error trap, sharing relevant experience and guidance before the job makes sense. In my experience, these pre-briefings are naturally happening in workgroups and settings as the jobs are distributed. For the person leading the pre-briefing, the challenge is to make it interesting and specific, and for the team member, it is about keeping an open mind and fighting back boredom from the seeming redundancy of it all. The closer to the job in time and space the pre-briefing is happening, the more effective it will be.

More important, however, is keeping an eye on the *competence*, *awareness*, and *compliance* of our fellow team members as we execute the task. That's at the heart of *teamwork*. We cannot take for granted that these tools will be utilized 100 percent of the time. Not for ourselves, nor for our co-workers. Blind spots, error traps for that matter, can be everywhere and can develop dynamically. As much as we can't guarantee for ourselves that we won't forget a detail, overlook something different as usual, or deviate from the rules, neither can our colleagues, we all have a responsibility to look out for each other.

That can imply calling out a more senior person on the wrong track, which is the whole point of crew resource management (CRM). CRM was developed after the power gradient between the captain and the first officer had too often prevented the crew from choosing the best option. CRM in aviation focuses on interpersonal communication, leadership, and decision-making in aircraft cockpits. After all, the captain won't always be right and sometimes the co-pilot must be the one to initiate the recovery. It doesn't only apply to the cockpit.

Jocko Willink, author of *Extreme Ownership: How U.S. Navy SEALs Lead and Win,* would call that leading *up and down* the chain of command. It's a skill and a mindset. It requires staff on every level to nudge their bosses in the right direction, fully encompassing the *teamwork* mindset.

With all that, we summarize up to here with the following:

TAKEAWAY NO. 4: We have a fighting chance to avoid error traps through competence, awareness, compliance, and teamwork.

The loss of the V2500 FCDs is a stark example of a design flaw that required twenty-six expensive incidents before it was determined that it was worth the time and effort to fix. Up until then, it constituted an error trap for all mechanics tasked to handle FCDs. This error trap was weakly defended, as the defenses only worked under certain circumstances, e.g., stepping back a few meters from the engine intake to see the latches or determine whether they were in danger of being forgotten or circumvented, like the logbook entry for every opening. We considered the new lockable latches as final fixes or strong defense to this error trap (yes, the key could be removed easily).

We will see that there is no shortage of these design-induced error traps, and it doesn't stop at design (or at aircraft maintenance, for that matter). Our best strategy to protect ourselves and others from these error traps is through the practice of *competence, awareness, compliance*, and *teamwork*. Those of us who work in positions of power over resources should keep in mind how easily small variations can lead to destroyed careers or even lost lives.

Those of us who lead people, especially the next generation of aircraft mechanics recruited from Gen Z, must think twice about what type of role model they must be. The young people struggle with building *competence* and developing *awareness*, making *compliance* and *teamwork* their life insurance while they build their skills and experience. Gen-Z's worldview is different from previous generations, and it is our obligation to better understand that. They may not have the same tolerance for persistent design flaws and hostile, non-forgiving environments that previous workforces had, and full *compliance* may not rank high on their list of priorities. Nevertheless, if we lead smart, we can teach them to be successful aircraft mechanics and engineers.

CHAPTER 2

Design Part I

AIRCRAFT AND PARTS

In our line of work, the primary responsibility of a leader is to ensure that all their staff are safe and will come to their respective homes complete and safe, the rest is secondary.

—CARL MICHAEL L. DE GUZMAN, BAY MANAGER

BUDAPEST, 2010. All hands were on deck for the arrival of the CEO and TechOps Chief of our newest customer, a small airline with a reputation for excellence. Their mercurial founder, chief pilot, and chairman was legendary for his attention to detail, and we were thrilled for the opportunity to maintain his fleet. This was our first event under this new partnership, and with an hour to go, I was preparing our small welcome ceremony. The hangar was spotless and the A320 was ready to be lowered from the jacks as the crew continued the C-check. Everything was picture-perfect. In fact, too good to be true.

BANG! The sound echoed through the hangar, stopping everyone in their tracks. The jacks' safety stay had punched through the tail of the aircraft. I looked at my watch, less than forty-five minutes until our newest customer arrived.

The Airbus aircraft from the A320 to the A380 are lifted on three points, with one forward jack and two wing jacks, shown below for an A321. In addition, a safety stay is used to prevent accidental movement of the aircraft, but it can't withstand any substantial load (only up to 2 of the 69 tos. max. weight), which rests on the three main jacks.

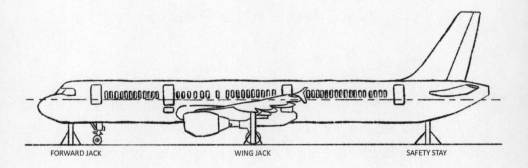

FORWARD JACK   WING JACK   SAFETY STAY

A321 jacking points. The weight of the aircraft rests on three main jacks (one forward, two wing), while the safety stay serves just for stabilization.

The safety stay must be removed before a retraction and extension test ("gear swing") and before lowering the aircraft from jacks. Numerous teams have learned the hard way of how easily the removal of the safety stay is forgotten before lowering the aircraft. It generates a distinct bang, when the tail cone structure is punctured with the safety stay in the way. Airbus is always standing by with repair kits, the huge doubler, additional structural elements, and repair teams if required.

Today, when I write about this incident, it doesn't seem to be a particularly original example; it's such an obvious error trap. An error trap with disastrous consequences. When you first see the stay sticking in the tail of the aircraft, you will be forgiven to wonder whether it's repairable at all. Most of the dents we deal with in aircraft maintenance are barely noticeable without a torchlight. Not this one. The aircraft fuselage isn't designed to absorb this kind of force from this angle and so the frames and stringers under the skin give way willingly. Good thing that it is indeed repairable and Airbus, here again, has a repair kit ready. The repair doubler for the skin is the size of a surfboard. It is expensive in all aspects, and as the icing on the cake, the patch looks ugly, especially for the Ferrari-driving owner of this forty-million-dollar baby on return to home base. My career couldn't fall much further.

Nowadays, Airbus prescribes the use of a dynamometer which provides an aural warning if the maximum load on the safety stay is exceeded. The technical aid to avoid this error trap seems to be effective, but of course only if it is consistently used and if the batteries are charged. Another danger comes from a removed safety stay which remains in the area, maybe because of the docking configuration, or when it is not lowered far enough. We have seen this scenario even when the aircraft was already on wheels but another task on the nose landing gear required the lifting of the forward jack, which pivoted the tail toward the parked safety stay. Obvious here, however, is the fact that this feature is so central to the design of this aircraft type that it cannot be changed.

A design can become an error trap if it is faulty, flawed, substandard, or problematic to maintain, as we have seen with the V2500 FCD. Often for the mechanic, there is not much more to do about it than to accept it as a fact and adapt to it as best as possible. Design changes are typically incredibly expensive and may only be feasible with the next generation, like the A321neo FCD warning. But sometimes there are technical means to the rescue which help avoid the error trap. Besides the aural warning, there are now also automatically synchronized jack sets available which lift and lower the aircraft without the need to manually control the leveling and the loads. As sophisticated and error-proof as this may sound, any additional technology may introduce new error pathways and may also increase the users' dependency on an automated solution—something pilots in modern aircraft must always be highly cognizant of.

For pilots, Automation Surprise occurs, for example, when the autopilot system behaves in a way that the pilots do not expect. An example of this is the Asiana Flight 214 in June of 2014, when the pilots relying on the autothrottle to maintain speed did not recognize that they chose the wrong autopilot "mode" for that, causing them to hit a seawall as they approached the runway.

"In their efforts to compensate for the unreliability of human performance, the designers of automated control systems have unwittingly created opportunities for new error types that can be even more serious than those they were seeking to avoid," Chris Hart, acting chair of the National Transportation Safety Board (NTSB), June 24, 2014.[7]

In-flight as well as on the ground, automated solutions, designed for higher efficiency and safety, may surprise the operators at times,

---

7   BBC News, "Asiana flight 214: Crew 'over-relied' on automation," June 24, 2014, accessed April 23, 2022, https://www.bbc.com/news/world-us-canada-28002054.

even when they are not malfunctioning, and even more so, if component failures come into the picture.

That brings us to:

TAKEAWAY NO. 5: Error traps from aircraft design can arise from design weaknesses but also from flawless design with problematic maintainability or automation features.

CAN ERROR TRAPS BE ELIMINATED?

"One of the defining characteristics of a safe organization is that it works hard to find and eliminate its error traps."[8]

Professor James Reason, one of the godfathers of human factors in aviation and beyond, doesn't say an organization can find and eliminate all error traps. He says it must work hard to do so, but is even that realistic? In the realm of aircraft design, that seems to me to be too idealistic. And what exactly does "eliminating" an error trap mean? Our first two examples have been stark error traps, which nowadays have been mitigated by technical means, but does that mean they are eliminated? And then we will see less dramatic error traps with less effective defenses. My point is that the aviation profession and our lives are full of error traps, making Reason's quote sound a bit quixotic. Sorry, Professor!

---

8   James Reason and Alan Hobbs, *Managing Maintenance Error: A Practical Guide* (London: CRC Press, 2017).

But all is not lost. Error traps can be significantly mitigated through effective defenses that are put in place to prevent the error itself. By definition, error traps have insufficient defenses, otherwise they would not exist. How insufficient?

We may look at two categories of error traps depending on their defenses:

Primary Error Traps

- *No defenses.* The defense of most technical issues that provoke errors is created only after the problem becomes apparent. For example, a warning note in the manual or a dynamometer with aural warning.

- *Weak defenses.* That's the default. A defense exists, but it doesn't provide perfect protection. Instead, it's, at best, a workaround to mitigate but not eliminate the error. A warning in the manual typically is a weak defense that is easily overlooked.

Secondary Error Traps

- *Clumsy defenses.* Then we have error traps with defenses which are not effective, because they hamper work, or they become a problem themselves. Long manuals with critical content buried under lots of clutter fall into this category.

- *Too many defenses.* And finally, for the same reason that makes defenses clumsy, too many defenses could achieve the opposite of what is intended. For example, multiple signatories for a single transaction can obscure responsibility.

Let's review the four tools we discussed in chapter 1 that will help you overcome insufficient defenses.

COMPETENCE: Master the basics. Reading and *understanding* the guidance helps to recognize the defenses for what they are. If you haven't heard of the type of incident the defense was put in place for, that doesn't mean the incident was trivial. With the tethered flag defense to address the FCD error trap, the mechanic didn't understand what the FCD key was meant to achieve even though its purpose was stated in the manual and would have significant consequences if not properly understood.

AWARENESS: Be in the moment, know what is coming. Vigilant *awareness* is especially essential when you work with simple or familiar tasks. When you "know" how to do a task, you may feel that the defenses reduce your efficiency and decide not to take them seriously. Here again, it is all about deciding consciously or unconsciously. Remember the mechanic back in Romania, in 2011, who told me that closing the latches on the FCD correctly is so important and delicate that he bumps them with the bottom of the fist to ensure that it is firmly closed—that's vigilant *awareness*.

COMPLIANCE: Respect the defenses. Reading and *following* the guidance means accepting the defenses and possibly the small penalty on efficiency. Following the existing guidance requires the least mental effort which could be beneficial if surprises arise. The manuals exist for a reason. Don't assume you know more than the hundred years of aviation knowledge and experience that the guidance in the manuals represents.

TEAMWORK: Looking out for each other lets you keep your eyes open to how your co-workers are handling the defenses. In the disaster with our new customer and the safety stay, if all team members had been communicating effectively and watching out for each other, the costly mistake could have been avoided.

In summary, we arrive at:

I am not saying "obey blindly." What I am saying is, take them seriously and don't work around them without first making a conscious decision based on knowledge, experience, awareness, and a proper risk assessment (we'll delve into the details of risk assessment in chapter 8).

MORE DESIGN CHALLENGES AHEAD

There are many more examples of aircraft design creating a maintenance challenge and where no technical aid to avoid the damage is on offer. For example, in removing the engine from an A330, all interconnections between the pylon and the engine must be uncoupled, one of which is the fuel line. The way the fuel travels from the wing tank to the combustion chamber of the engine, as depicted below, is through a low pressure (LP) fuel shut-off valve to the fuel oil heat exchanger, further to the fuel pump and through high pressure (HP) to the fuel metering unit (FMU) and the fuel nozzles.

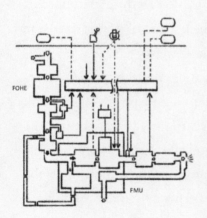

A330 fuel system (engine side). The circuit breakers controlling the LP fuel shut-off valve (upstream in the pylon) and the HP fuel valve (in the FMU) have to be opened in the right sequence. The fuel is pumped from the upper left thru the fuel oil heat exchanger (FOHE) and the fuel metering unit (FMU) to the lower right corner in the illustration.

Before disconnecting the LP fuel line from the pylon, the remaining fuel must be drained. Before opening the drain plug at the heat exchanger, the mechanic must pull three circuit breakers in the right sequence: two for the LP fuel shut-off valve actuators and one for the HP fuel valve in the FMU. If the mechanic mistakenly deenergizes the HP valve first, the LP valve fail-safe logic will cause it to open. In this case, the mechanic would be showered with fuel as it flowed from the wing tank to the drain port. This is an example of another "automation surprise."

While extremely uncomfortable, it is not catastrophic as they will be able to reinstall the drain plug immediately (the pressure is low, so the plug can be installed against the pressure), stop the fuel, and correct the error.

This brings a more extreme scenario to mind which is well remembered by older colleagues: A leak of the A330 trim tank fuel line between the center tank and the trimmable horizontal stabilizer (THS) made it necessary to disconnect it midway. The fuel line is embedded in a shroud to prevent leaking fuel from penetrating the cargo compartment. This defense makes it harder to find the leak. Part of the procedure is to defuel the trim tank. The manual says: Step 1—Defuel the THS trim tank, Step 2—Drain the remaining fuel from the tank.

Defueling the tank seems to be common sense, so it isn't even considered worth a warning note. But in at least one instance, the mechanic thought that the valves (the trim tank isolation valve and the trim tank pipe isolation valve) were already closed when he approached his work, leading him to believe that the remaining fuel in the pipe was not under pressure. After all, no fuel was flowing from and to the trim tank as he started the work. Difficult to know the exact configuration in hindsight, but he didn't think much about it, as he was the one to defuel the aircraft the day before and the trim tank was empty. Until it wasn't, as the night crew had transferred fuel to perform a fuel leak check on the front spar of the trim tank.

If this system behavior or the configuration isn't fully clear to the user (read and understand), just going by the manual would have been an easy way out (read and follow). It didn't happen. The mechanic opened the trim fuel line and the fuel from the trim tank started to flow into the drain container, *all the fuel* in a firehose-fashion, all the four tonnes from the trim tank, unstoppable due to gravity. Combustible fuel was everywhere inside the aircraft and on the apron. Non-compliance of the existing defenses created a costly and potentially dangerous situation.

Here's another one. A very basic one.

The pitot probes sit outside the fuselage facing forward. The air pressure flowing into the tubes when the plane is in motion enables the system to compute the aircraft's velocity. If the aircraft is on ground for longer, covers are put on the probes to preclude contamination. Now, because they sit outside the fuselage and are exposed to very deep temperatures, they have heaters. When a plane is simply being repositioned, it is relatively common that the covers are forgotten and, as a result, quickly burn once the aircraft is under power requiring the whole port to be replaced. By design, the covers are required, but there is no technical aid that will alert the mechanic to remove it.

B777 wing. The leading-edge slats can collide with the thrust reverser (shaded triangle).

Another good example is the flap/slat system of the B777. Sometimes the flaps and slats must be positioned independently from each other in maintenance. With the flaps manually extended, the slats could unwantedly extend once hydraulic power is provided again, if not properly deactivated. And then the slats may hit the open thrust reverser. This is interesting as it is not obvious which system would move the slats and it is different from that on Airbus. The mechanics might assume, after verifying that the flap lever matches the actual slat position, that the slats can't move, even if the system is not fully deactivated by pulling the circuit breakers (and for additional safety also removing the plugs). But it seems that in an unexpected configuration such as this one, plus perhaps a small component failure, the slats can indeed extent, even with the lever in "up". If the procedure is followed, deactivation of the slats in this case, safety will be assured. If not, surprises such as unwanted movements can happen, and it will be very hard to determine whether the lever was in "up" or not.

That the slats could hit the thrust reversers is a geometrical problem. The slats extension and the thrust reverser opening interfere with each other (see shaded area in the figure). It is made more challenging because multiple teams often work on the engine and on the flight controls simultaneously. An inherent error trap to be sure. Here are Airbus' and Boeing's warning notes:

The Airbus manual says, "If the thrust reverser cowls are open, make sure that the half doors of the thrust reverser are closed before you extend the slats. if the doors are open, you can cause damage to the slats and half doors when the slats extend."

And the Boeing manual says, "WARNING. Make sure the thrust reverser cowls are closed when you operate the flaps and slats. The slats can hit the thrust reverser cowls if they are open. This can cause injuries to persons or damage to equipment."

Again, *competence*—system knowledge—would be the best way to ensure against unwanted movements. *Awareness* would put you on high alert when switching on power again after maintenance has already started. *Compliance* would force you to complete the deactivation procedure step by step and double-check when in doubt. And *teamwork*, finally, would require just one person to stop the rushed action.

One more example: The installation of the V2500 no. 5 bearing compartment cover. If the seal (or packing in the drawing) is not seated perfectly after torquing the thirteen bolts, an oil loss may occur. This is basic in aircraft maintenance. There are so many seals on the aircraft (also called gaskets or packings) which must be well seated, otherwise leaks will occur. There are different techniques to make sure that the seal is well in place. One is visual with mirror and torchlight. Another one is torquing in steps.

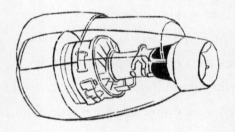

Location of V2500 no.5 bearing.

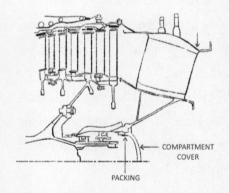

COMPARTMENT COVER

PACKING

V2500 no. 5 bearing compartment cover and packing. The correct seating of the packing is critical.

These are just a few examples where the aircraft design makes maintenance an expert's trade. Manuals contain thousands of warning and caution notes, which in many cases, came from experience and incidents. A simple task like topping up oil for the A330 Green Hydraulic Reservoir has an eight-page set of instructions filled with warnings and cautions. Some are trivial, like be careful not to get fluid in your eyes. Others not so much as with the reset of circuit breaker 5JR for the HSMU, which might lead to an air-turnback if forgotten.

AIRCRAFT PARTS HAVE THEIR OWN DESIGN FLAWS AND WEAKNESSES

Let's turn to aircraft parts. The installation of the right parts is a critical component of the airworthiness of the aircraft. We have seen accidents and deaths due to the wrong parts used. And there are many ways to get it wrong. We look here at design first. Which parts are designed in a way that lay an error trap for maintenance?

Here we start with parts that fit in two or more different ways, with only one way to be correct. Let's look at the A320 cargo door seal. It contains inflation holes that let in air from the pressurized cargo hold in flight. The inflated seal will optimally prevent air leaks after pressurization. The problem, however, is that it can also

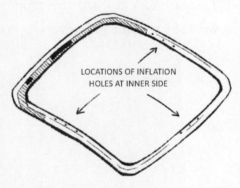

LOCATIONS OF INFLATION HOLES AT INNER SIDE

A320 cargo door seal. The nine inflation holes shall point to the inner side.

easily be installed improperly with the holes on the outer side. When placed with the holes on the outer side, the cabin cannot pressurize after takeoff and the flight will be forced to return to its origin.

The defense is the instruction and the caution note in the manual. That's it. Because we are challenged to read infinite caution notes, one that says, "make sure the seal is installed in the correct position," can be easily overlooked resulting in a weak defense.

Another possibility is to mix seemingly interchangeable parts between their positions in the aircraft, which is why meticulous parts tagging is so important. The links of the A380 movable fairings can also be installed in the incorrect position. When that occurs, their lengths will be incorrect when extended, which can damage the flap track fairing.

All of these are examples of error traps originating from design. What makes design special here is that it is so hard, if not impossible to change, especially in the short term and especially from the perspective of one user or one organization. "Identify and eliminate" your error traps is indeed a tall order. Instead, we must adjust and adapt to them as a fact of life. These error trap examples are just the beginning. There is more at work than problematic designs for maintenance, as we will see in the following chapters.

CHAPTER 3

Design Part II

TOOLS AND PROCESSES

Shortcuts are borne out of previous
shortcuts and only lead to disasters.
—EDGAR ALLAN FARIÑAS, TEAM LEADER

MANILA, 2006. We were in a festive mood. It's Friday evening and we have just received our first ten-years' check (which we formerly called D-check) from the Middle East. The captain left us with the words, "This aircraft needs an overhaul." And, indeed, it turned out to be in bad shape requiring over hundred parts replacements on the floor structure—something we had not experienced before. But that's OK, that's our job.

On day two of the check, we hosted our guests, the customer's technical representatives and their primary base maintenance and engineering partner, for dinner, in Bonifacio Global City. The primary partner is a competitor of ours and the check was originally scheduled to happen at their facility. For one reason or another, that wasn't possible, so here we sat with our new friends and our competitors over Italian dishes and wine. It felt a bit strange, but things could be worse, and I would soon find out just how much.

Just as the entrée was being served, I received a call from the hangar. "We have a problem, Boss," said my crew chief. "A big one."

I quickly excused myself and stepped outside to hear the news in private.

"We dropped an engine."

"You dropped what?"

The engine had toppled over while hanging on the four bootstrap chains midway between the 6.50 m high pylon of the aircraft on jacks and the ground. The nose cowl almost touched the ground and the exhaust plug hit the pylon. The rest of the engine in between had made contact with the chains which damaged the fan case. An unforgettable image to be sure.

Removing all engines and pylons at that time was a top priority as some pylon parts had to be sent for rework to Toulouse and back, which was always very tight for layovers no longer than thirty days. The rest of the evening was spent salvaging the toppled engine to keep it from falling to the ground.

So, what happened? This was the first time that we had used a new model of an engine dolly or "in-shop trailer." It was much lighter than the older model and thus much easier to maneuver, especially hoisting it up on the bootstrap chains, which was done manually at the time. It was that improvement of being lighter that became a problem when the dolly was lowered with the engine. The process requires that all four sides be lowered in perfect synchronization so that the engine stays level.

WARNING: MAKE SURE THAT THE ENGINE AND TOOL ASSEMBLY IS ALWAYS LEVEL AND NOT IN A NOSE-DOWN ATTITUDE. IF YOU DO NOT OBEY THIS INSTRUCTION, THE ASSEMBLY CAN FALL AND CAUSE DEATH OR INJURY TO PERSONS.

If the center of gravity moves forward, it may become unstable, especially when the nose cowl is still installed. The lighter dolly was not forgiving of any imbalance, and so the engine toppled.

This new dolly had certain restrictions of use:

Right use:

- Trailer is used to carry in-shop the *CFM56-5C* engine and to make the bootstrap installation.

- You can carry the trailer with suitable trucks using the towbar 537201193 or equivalent bars.

In-shop trailer for CFM56-5C. The illustration shows the light model, which is easier to maneuver but more prone to instability.

- The traction eye needs a minimum ground space of 12 cm.

- You can only move the trailer on solid and plain streets.

- Max towing speed is 5 km/h (3 mph).

- You can only use the 2. handbar for maneuvering.

Foreseen misuse:

- Over-speeding and driving on rough ground.

- Trailer with installed engine carried by a forklift.

- Installing handbar during towing.

- Towing without blocking the 2. Steering axis.

- Trying to install an engine type other than the CFM56-5.

- Not locking on the lash-points during air transport.

It can be used to install an engine using bootstrap, and it has provisions for that labeled "bootstrap." The term "removal," however, wasn't mentioned, but also not included in the misuse list. So, it wasn't obvious and not even logical, that you could install the engine that way but not remove it. It is exactly the same, just the sequence of steps in reverse.

The balance problem would be there either way. But strictly speaking, this piece of equipment was not properly authorized for removal of the engine. It looked legitimate, but it wasn't.

Tools may become error traps. The criticality of keeping the engine level when hoisting doesn't go away, but it can be alleviated by a counterweight on the cradle. That extra weight creates a greater strain when pulling it up, but it is much safer. And safety can further be enhanced by an electrical hoisting system.

Apart from those defenses, the four tools discussed above will also help avoid the problem. *Competence* will inform you that keeping the engine level during hoisting is crucial, it should be part of a pre-briefing for the team, including instructions for those four who control the hoist blocks and the commands to be used.

Awareness and alertness should be straightforward for an obviously dangerous task like this one. *Compliance*, here, means taking all the warnings and notes seriously and double-checking that all tools and equipment are the correct ones and in good condition. If there are any means to check or calibrate the equipment before the task, use them. And, finally, *teamwork* includes watching out for crew members who are not fully qualified, not experienced, or not at their best today.

SPECIAL TOOLS: A HIGHER POTENTIAL FOR ERROR TRAPS

All the challenges related to aircraft design leading to error traps in maintenance have the potential to be more pronounced for special tools which have a small market, high development and qualification costs, and are sparsely tested.

The ground support equipment (GSE) tool used in removing the A380 THS actuator falls into this category. To remove the A380 THS actuator, the mechanic must take the weight of the THS or horizontal tail plane (HTP) from the actuator so that the bolts can be removed. For that, Airbus prescribes a special bar tool, which transfers the weight of the THS to the airframe structure at Frame 101. This tool has a linear actuator to extend and retract hydraulically operated by a small pump. The tool is installed at Frame 101 and extends the shaft until it reaches the THS. The tool is then coupled to the THS and applies a load downward to take the weight of the THS from the actuator. The challenge here, however, is to know which direction is which on the pump. The lever has no markings and once installed the mechanic won't see anything moving. If the piston is moved in the wrong direction (extension) the bolts won't be freed up. Worse, a force is applied to Frame 101 that it is not designed to take. It may crack, resulting in a massive repair task.

The manual warns, "Caution: Before you install/remove the GSE-tool, read the instructions in the applicable GSE instruction manual. While you operate this tool, carefully obey them. If you do

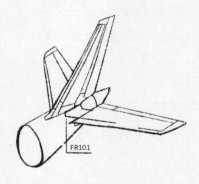

Frame 101 ("FR101") at the A380 tail section. This location serves as the aft jacking point and also the attachment point for the bar tool used for the THS actuator removal.

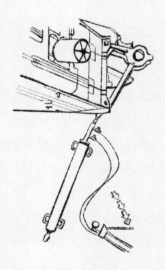

A380 bar tool. Transfers the weight of the THS to the aircraft structure at Frame 101 in order to remove the THS actuator fitting. The tool's piston is extended and retracted by a hydraulic hand pump (lower right).

not obey the instructions, damage to equipment can occur."

Meanwhile, Airbus has removed details about using the tool from the Aircraft Maintenance Manual (AMM), which formerly made it tempting to ignore the more detailed "(GSE) INSTRUCTION MANUAL." The problem remains, that you don't know in which direction the piston moves, unless you take note of it while the tool is not fully installed yet.

The instruction in the GSE manual is to "use the hand pump in retraction." The lever on the pump has no label, but since it has just been used in extension to connect the tool to the THS, the mechanic must switch the lever to the other side and transfer the load up to 16.69 kN (or 1,693 kg-force). The maximum load for the aft safety stay point, which is also at Frame 101, is 4,500 kg-force, and both load limits apply to inward force. The limit of maximum outward force is much smaller, so the manual warns, "Do not apply any compression load." But this is not very intuitive, as the mechanic doesn't know whether the

HTP is pulling upward or pushing down (the pivot is further to the aft) and they can't see anything moving once the tool is connected.

So the mechanics indeed mixed up the lever position, reached 16.67 kN but couldn't remove the bolts. That was their last chance to notice their mistake. They tried again and Frame 101 cracked. Massive repair. Nowadays we have placed a label on the case.

PROCESS DESIGN CHALLENGES

Finally, in addition to tool design challenges, we must also address potential error traps inherent in the design of processes. Solutions for hardware design problems, if a solution even exists, are often very expensive. But if there is a fail-safe design, it is then typically difficult to circumvent. With processes, it is in a sense, the other way around: they can be, at least in theory, relatively easy to change if the organization has an efficient process to do so. But then they are also often victims of optimization at the hands of the users.

As an example, consider the parking procedures which during the pandemic, came to prominence on a scale like never before. With 50–90 percent of fleets grounded, line maintenance teams around the world had to deal with long-term parking and storage processes they were often not familiar with. In normal times, an aircraft on ground (AOG) is a highly exceptional condition. Parking is seldom, let alone for long term, outside of dedicated facilities typically in the desert. The scheduling process for line maintenance is typically based on the next flight. Tasks are due before the next flight. If the task is due today, but the next flight is tomorrow, the task is effectively due tomorrow before the flight.

Priorities are constantly shifted around to complete tasks on aircraft due for flying and remove tasks from aircraft that will remain in the hangar. That process, however, doesn't work with parking. A run-up, also known as an engine ground run (EGR), which is due within thirty days, must be performed in time even without a next flight. Many line maintenance systems had to be adjusted, partially by manual control, to keep on top of the parking requirements in an otherwise highly automated system.

Here, the process hindered our ability to efficiently adapt. That might be an exception nowadays, as many organizations try hard to streamline movements and activities. Contradictory or nonsensical steps are typically found and eliminated. But it is different for additional process elements that are designed for safety.

Processes often add redundancy to safety-critical tasks to make them more fail-safe. That could be an independent inspection (or "buy-back"), rotation of performers (like for Extended-range Twin-engine Operational Performance Standards (ETOPS) tasks), additional documentation requirements (like the techlog entry in the FCD case above), or a checklist, which summarizes again what should have been performed already. Redundant steps, however, can become error traps, when it seems obvious to the user that they are—well—redundant, so that the temptation to bypass them in the interest of efficiency may be strong. For the users, the difference between an important element of safety and a nonsensical step might not be straightforward at all.

That leads us to:

TAKEAWAY NO. 7: Users may choose efficiency over compliance with process redundancies that are intended to ensure safety.

The engine incident, at least, had some fortunate outcomes. Impossible as it seemed, we were able to keep the thirty days turn-around time (TAT) for the layover and the customer was able to return his aircraft to service as planned. CFM56-5C engines were hard to find at that time, but miraculously, our planning and engineering chief had good contacts in Sri Lanka where he grew up, and we found a spare engine there for loan. We were able to conclude a commercial agreement with the Sri Lankan carrier within days and brought the loan engine to Manila. All the inspections and repairs on the aircraft side, especially on the pylon, could also be completed. This aircraft needed an overhaul, and it got one. On time. Our customer was, of course, not happy with the incident, but they were massively impressed with the recovery. Even our competitor on-site was impressed. The damaged engine was sent to the shop, repaired, and back in the plane two months later and the borrowed engine was returned.

There exists an unlimited number of error traps originating from design of the aircraft, its parts, tools, and processes that require the consistent employment of *competence, awareness, compliance,* and *teamwork* to mitigate these traps to the best of our abilities. Now that we have looked at how these error traps present themselves in design, let's delve into the error traps that arise from human factors.

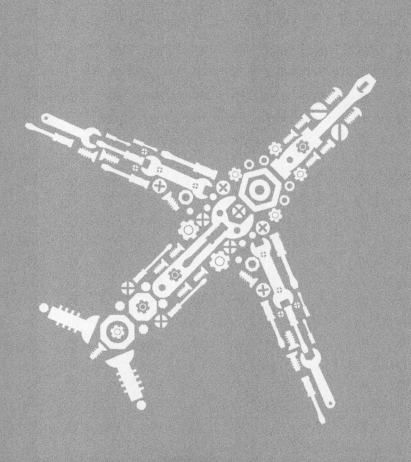

CHAPTER 4

Human Factors

BASIC AND ADVANCED

Studying human factors does not equate to advocating for the blaming of workers as the prime cause of accidents [...]. While this might be unfashionable to state, there is a level of personal accountability for worksite safety in almost every occupation.

—RHONA FLIN, IN: *SAFETY SCIENCE RESEARCH: EVOLUTION, CHALLENGES AND NEW DIRECTIONS* BY JEAN-CHRISTOPHE LE COZE

HAMBURG, 2005. An emergency inspection came in from Airbus while one of Lufthansa's A340s underwent its first ten-years check. We had to examine a connector in the fuel tank which was meant to be a production improvement. This connector ties a fuel pipe feeding through a hole in rib 6 to the aircraft structure. Rib 6 is the part of the wing which also carries the main landing gear. This connector introduced stress to the wing rib causing cracks to develop from its bore holes. This type of crack—a huge one—was found by chance on another aircraft on a routine inspection in Dusseldorf earlier that year. Cracks in the aircraft structure can be dangerous for the integrity of the airframe in flight, but they usually develop slowly. How slowly depends on multiple factors, and it is often hard to estimate. In many areas that are exposed to stress, we look for still invisible cracks with methods of non-destructive testing (NDT). When the crack becomes visible it has been there for some time already, and it may take years to grow into concern that it may weaken or even break the structure.

Now that we knew this type of crack existed, we easily spotted the crack on the A340, sometimes even without the help of NDT. The significance of this crack extended the layover by many weeks as Airbus worked out a repair. After that occurrence, the same crack was discovered on almost every affected aircraft. Why is that? These cracks were not new. They had simply been overlooked for some time. The first one found had already propagated "through the full thickness of rib 6" before being detected. Once you know where, or better, what to look for, they become visible. Psychologists call that "expectation bias." You find what you are looking for, but you easily overlook everything else, so that afterward people would ask, "How could you have missed that?"

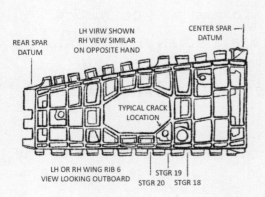

A340 Rib 6 typical crack location. This is a cross-section of the wing showing the location of the possible crack.

"An A330 operator has reported during a maintenance check, significant cracking of LH and RH wing rib 6 aft web. The cracks are located in the lower part of rib 6 aft aperture, between bottom skin stringers 18 and 20, and extend from the lower edge of aperture in rib 6 to a fastener hole and then into the fuel pipe hole. The crack has developed through the full thickness of the rib 6." (AD No F-2006-009)

This wasn't the first time a crack finding by chance resulted in the discovery of many similar findings which were formerly missed. Fatigue cracks in the fastener holes of the skin lap joints of an aging B737 were behind the accident of Aloha Airlines Flight 243 in 1988. The aircraft lost a large piece of its fuselage skin in flight but was able to land safely. The cracks should have been found by regular maintenance inspections, especially as there were specific Boeing instructions out for lap joint inspections including an Airworthiness Directive (AD), which was performed on this aircraft the year before the accident in a different area of the fuselage. The dislodged piece was never found, and the examination of the remaining structure revealed many visible and even more invisible cracks in the area of the inspection triggered by the AD. In addition, even a passenger saw a crack upon boarding that fateful flight but failed to report it immediately.

The National Transportation Safety Board report identifies, among other factors, the low probability of findings as a cause for the absence of these findings in two ways. First, the original D-check structural inspections, which normally take place in a three-to-five-week hangar visit, were atomized to fifty-two blocks that were often performed during night checks. These night checks happen under the pressure of the next scheduled flight in the morning which any structural findings could easily jeopardize. So, inspectors would already be inclined to consider these inspections a formality with certainly no findings, otherwise flights would be cancelled in the morning. It shouldn't be like this, but we have seen it. Put an all-cabin seat belt inspection (for fraying of the material) on a night check and you may get zero findings. The same inspection on a three-weeks C-check returns fifty seat belts for replacement.

Secondly, the report points to the fact that the AD inspection for example, included the close visual inspection of 1,300 rivets in a straining body posture. Here it becomes hard to stay focused on the job, which is boringly uniform and exhausting at the same time. Both factors could lead to missing a crack due to the "expectation of results": nil findings.

Expectation bias lets us see what we expect to see and ignore the rest. It is not only present when inspecting hundreds and thousands of rivets. It is a tendency of the human mind every inspector must deal with every day. Your previous findings or absence of findings can and will influence the way you look for deviations in the future. You develop focus areas and blind spots. That's why inspectors should rotate from time to time, as should auditors, as should managers. Any other counterstrategy? Not easy because as with fatigue, we are generally not good at assessing our own limitations in real time.

Expectation bias is a rather unusual starting point for a discussion of human factors. But as we now find ourselves at the "advanced" end of the spectrum, let's further look around.

CONFIRMATION BIAS

In addition to expectation bias, there is confirmation bias, which can also get in the way of good judgment. Closely related to seeing what we expect to see is seeing what we like to see, what confirms our view, our prior judgment. We like to see facts that embolden our beliefs, and we disregard opposing inputs. This could become a problem in troubleshooting, especially in an emergency. If we make up our mind, we are often uncomfortable to change it. And then we ignore facts that contradict our initial judgment.

Confirmation bias can also be present in our relationships and preferences. We like or dislike certain people, hotels, restaurants, and airlines, and we seek out details that support our earlier impression. We have customers who like one bay, one team, much better than the other, and then they find proof to support their preference. Well, it may be true that one team performs better than the other, it may also be that we are sorting the available bits of information and impressions to fit our perspective. Sometimes people dislike places for reasons they can't even remember.

More importantly, confirmation bias may hinder your consideration of unpleasant scenarios. When you don't find a fault, you must start challenging your assumptions layer by layer. For those of us who have spent time with software programming, this sounds very natural, because debugging programs is such an integral part of the job, and because we have all seen bugs that were so persistent that we had to go back to the most basic assumptions and literally trace 0s and 1s and no longer take anything for certain. Mark Twain said, "It ain't what you don't know that gets you into trouble. It's what you know for sure that just ain't so." In other words, our assumptions are indeed important in life and our minds protect them against contradicting information, which, unfortunately, sometimes turn out to be wrong.

The Big Short, which opened with the above quote from Twain, is a film about the financial crisis in 2008. The film shattered the assumption that all participants in the economy behave objectively in line with their best interests, so that all decisions are rational. This paradigm was finally dislodged by the school of Behavioral Economics spearheaded by the work of Tversky and Kahneman, but it took decades to be accepted into the mainstream. Richard Thaler (who also appeared in the movie) was lingering in obscurity in the '70s and '80s, before he got a call to Chicago's renowned university in 1995 and went on to win the Nobel Prize in 2017.

And even today we would find people who don't fully buy into this concept, and prefer to consider those errors, those failures of judgment, as random or obscure, or maybe even superstition. We are familiar with the saying that the first impression is the best impression. Most audits we go through end badly when they start badly and the other way around. Why should that be true? Is it really that, when the auditor encounters some bad housekeeping on day one, we will see more non-compliances in the audit report than expected? Or is it superstition? For now, let's assume that is real and if not, we err on the safe side.

In the practice of aircraft maintenance, expectation bias and confirmation bias may hamper optimal inspections and faultfinding, because we see what we expect to see, or we see what we want to see. Part of our *competence* should be to understand those particularities of our mind, which are not deficiencies per se but help us in many other situations. Our *awareness*, our special alertness, should kick in when we encounter situations that are prone for bad decisions due to our biases, because our mind plays tricks on us. We sometimes must return to our assumptions and our preferences and challenge them. This might, of course, be wishful thinking, especially under time pressure.

Time pressure is the biggest enemy here and the very reason why our brain uses these heuristic techniques in the first place. These are nothing more than mental shortcuts to save time and energy for more important tasks. We are often unaware that we are making these mental shortcuts in real time. The psychologist Marshall Goldsmith recommends that we use some *structure* to keep us in line with our good intentions. He, in fact, asserts that we cannot get better without it. The structure can be established with the help of an attention-grabber like some hardware or a recurring event that reminds us of our original plan. All of us, without the aid of structure, can become easily distracted and forget our original intentions.

Our third line of defense is *compliance*. If we acknowledge that our brain loves shortcuts, because at one point it assured survival in the face of saber-tooth tigers, and that these shortcuts can work against us when they are based on the wrong assumptions, we should count deliberate *compliance* as an ally. Every time we go by the book, we make it harder for wrong assumptions to influence our work.

And finally, *teamwork* should always provide additional protection. Suboptimal decision-making under the influence of those mental biases, or other human factors for that matter, is hard to avoid solely by our own good intentions. We can, however, more easily detect it in others. Maybe even easier than you would objectively expect, that's due to another human flaw.

Plan-continuation bias, or what pilots call "get-there-itis." This is another interesting trick of our mind. We go ahead with our predetermined course no matter what and ignore new information. Pilots who land the plane as planned rather than divert despite the severely deteriorating weather are a good example.

These three mental biases are just examples of many more similar decision-making deficiencies, which let us choose the wrong option, especially when the decision is tight between the alternatives. In such cases, our mind is conditioned to shortcut a lengthy elaboration of benefits and risks and go by heuristic techniques or rules-of-thumb. In other words:

> **TAKEAWAY NO. 8:** Mental biases are shortcuts of our mind. Sometimes they are based on wrong assumptions that can lead to unfavorable decisions.

So, this is the far end of the spectrum of human factors. The near end, which probably is also part of my esteemed reader's mandatory recurrent training and the subject of very good textbooks: The classic Dirty Dozen:

1. Lack of communication

2. Distraction

3. Lack of resources

4. Stress

5. Complacency

6. Lack of teamwork

7. Pressure

8. Lack of awareness

9. Lack of knowledge

10. Fatigue

11. Lack of assertiveness

12. Norms

These basic human factors are often used to explain why things went wrong. *Awareness* of these is considered indispensable for aviation maintenance management and staff, and that's why it is included in most recurrent training agendas. Without repeating the content of these trainings, I'll highlight what stands out to me.

"Pressure" stands out because it is an underappreciated fact that belongs at the top of this list. One of the excellent textbooks on human factors, published by the Australian Civil Aviation Safety Authority, states, "Few industries have more constant pressure to see a task completed than aviation maintenance." In our world, pressure is as certain as death and taxes, and with that certainty comes consequences. Almost all mechanics involved in incidents claim "time pressure" to be a contributing factor. If you agree to the above "certainty" of pressure, then your *competence* should not allow you to explain something unusual, the incident, by something constant, time pressure. In theory at least. In the real world, how pressure impacts the risk of errors often depends on the leader on the ground, including aggravating or alleviating time pressure in the team.

Unfortunately, some mechanics understand those basic human factors only on a very superficial level and even consider them a reasonable menu of excuses. It is indeed a list of error traps, and there is only so much that the system can do to eliminate them. This is again where *competence*, *awareness*, *compliance*, and *teamwork* should be helpful tools to mitigate them in daily work.

Between the near end and the far end, there are several more specific human factors that warrant attention. Tony Kern in his book, *Blue Threat Field Book and Study Guide*, calls them "Error-Producing Conditions" (EPCs), "Violation-Producing Conditions" (VPCs), and "Hazardous Attitudes" (HAZATs).

We list these below as we find them in Kern's *Blue Threat*:

EPCs:

1. Fatigue/Physiological Degrade

2. High-Risk/Low Frequency (HR/LF) Events

3. Time Pressure

4. Low Signal to Noise Ratio (LSNR)

5. Normalization of Deviance (NOD)

6. One-way Decision Gates

7. Information Overload

8. Poor Communication/ Information Transfer

9. Faulty Risk Perception

10. Inadequate Standards

11. Distraction

12. Broken Habit Pattern

13. The Thirteenth Offender: The First Mistake

VPCs:

1. Mission Expectation

2. Ego and Power

3. Unlikely Detection

4. Poor Planning

5. Leadership Gap

6. Poor Role Models

7. Unique Event

HAZATs:

1. Anti-authority

2. Impulsiveness

3. Invulnerability

4. Too Competitive/Macho

5. Resignation

6. Pressing Too Far

7. Oversized Ego/Vanity

8. Emotional Jetlag

9. Along for a ride

10. Procrastination

This is just one way to look at it, and this list is not exhaustive. There are overlaps and repetitions with the original list of the Dirty Dozen. Every entry could be used to spark a discussion in an advanced human factors class. Just to pick one: "The Thirteenth Offender: The First Mistake" is an interesting one, which is not easily assembled from the basic human factors. Often our first mistake triggers an overreaction, for example out of reflex or panic, like in "Never catch a falling knife" (in the broader sense, not in the stock market meaning). And often these second errors are far worse than the first ones. The worst of all is to "cover up" the first mistake, which can turn a triviality into a crime.

To conclude:

TAKEAWAY NO. 9: All human factors can be error traps. Our competence should go beyond the Dirty Dozen.

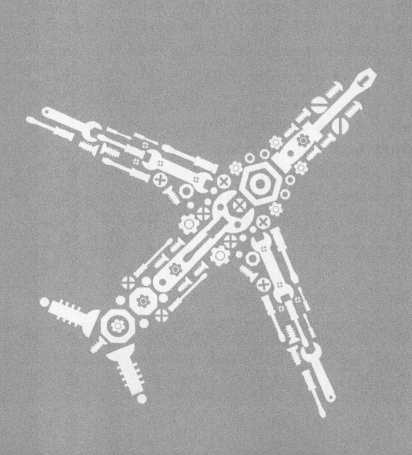

CHAPTER 5

Even Manuals Can Be Wrong

HOW WE PERPETUATE PAST ERRORS

*If you do not know something, ask
about it; if you know, teach it.*

—JAYSAR B. DINO, FOREMAN

MANILA, 2007. We were working on a routine C-check on a Philippine Airlines A340. One of the four engines had just been re-installed when one of our inspectors asked for proof that the torquing was done correctly. Four bolts each connect the engine to the pylon on the forward and on the aft side. The manual instructs to torque these bolts from the nut side at the forward mount (the aft side has anchor nuts, so it's only possible to torque from the bolt side). From the nut side, that means torquing from underneath where there is minimal space for the massive torque wrench. So, our people used a rule of thumb of applying 10 percent more torque than prescribed to do it from the bolt side. A rule they said they took from standard practices (ATA Chapter 20), or was it a Boeing rule? In either case, these rules of thumb do exist for some cases, but in this case, we couldn't find any allowable alternate way to do this job. It simply didn't exist. Next, ask Airbus.

Airbus gave us a "no technical objection" (NTO) statement to torque from the bolt side and the case seemed to be closed. Not with our inspector. He argued that an NTO does not constitute an acceptable piece of approved data, and that we needed to ask for a proper "technical adaptation" (TA). The game of chicken began. Airbus and their supplier Goodrich were not amused that fourteen years after entry into service, some customers would ask to change a procedure that was in use worldwide. They referred us back to the manual and encouraged us to use standard extensions for the torque wrench. Our engineering chief was working day and night but couldn't find anything that would get the job done on at least three of the four bolts. Airbus called up a French airline. They confirmed they worked as per the manual. I called up Frankfurt. They confirmed it is not possible to torque from the nut side. We asked for photos, videos, or anything as proof from Airbus that it could, in fact, be done per the manual's instructions.

Meanwhile, our customer grew extremely nervous at the possibility that we had mis-torqued all engines on their A340 fleet. They called their friends at other airlines and were told that everybody was working according to the manual. Shortly before this game of chicken ended in a head-on collision and shortly before the whole A340 world fleet faced grounding, Airbus called their French customers again. This time, they heard what we knew already. A way to torque from underneath did not exist. The charade quickly unraveled. Airbus gave in. Goodrich gave in. They sent us our TA. They confirmed the torque values (interestingly, nominal values also from the bolt side). They changed the manual. For fourteen years, everybody had done it incorrectly, and those who applied that 10-percent rule of thumb had even over-torqued their engine bolts. Nobody looked great in this episode, including us. Why only now was the truth confronted?

Errors may live on if not detected and corrected immediately. Once an incorrect performance has slipped through all defenses without apparent adverse effects, it's likely to be perpetuated. It is hard to flag a deviation the second time around, when the first time went fine. Even harder the tenth time, one-hundredth time, or after fourteen years.

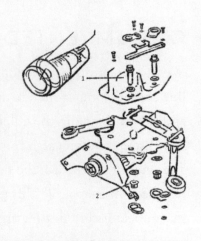

A340 CFM56-5C forward engine mount with attachment bolts. There are four bolts (1) secured by nuts (2).

"How did you do it last time?" "Everybody else can do it, why can't you?" is embarrassing to hear, but that's still better than non-detection which is more likely. That's not a good thing, but we are so used to these types of small adaptations, and without apparent adverse effects, it has the "normative force of the factual." Adverse effects maybe far in the future, maybe small and indistinguishable from other factors, or may not happen at all like in the case above. The adaptation becomes the new official procedure.

Adaptations without any lasting adverse effect may well fly under the radar for a long time. If defenses serve as additional means to work safely, mechanics may consciously or inad-

Torquing of the A340 CFM56-5C forward engine mounting bolts from the upper (bolt) side holding the nut on the lower side.

vertently bypass them. If nothing goes wrong, the deviation may not only be inconsequential but also may not leave any trace, and so may go undetected for years. Let's look at an example.

CIRCUMVENTED DEFENSES

We had to perform a functional check of the A330 trim tank transfer line shroud connectors. The transfer of fuel between the main tanks and the trim tank in the tailplane requires a fuel line running all the way through the aft cargo compartment. To safely collect leaking fuel and drain it overboard, the fuel line is embedded in a shroud, as we have seen in chapter 2. This test is to make sure that the shroud and its connectors are not leaking themselves, so gentle air pressure is applied from inside and the test is passed when the shroud can hold the pressure. This air pressure is just 15 psi. By some unfortunate circumstances, the mechanic applied a much higher pressure without noticing it, and the whole fuel line collapsed and the shroud burst. The striking point here, however, was the prescribed pressure relief valve that would have limited the overpressure to 20 psi. "Would have" if he had used one and, in this case, made worse by the fact that we didn't have one in the tool room and never had.

Relief valve. Standard industrial model which can be adjusted to open at a specified overpressure.

The relief valve is a classic defense, making it impossible to apply unsafe pressure. But it's only a successful defense if it's utilized. The manual is clear, and the valve is cheap, off-the-shelf, and commercially available. There is no excuse not to have it. Sometimes relief valves are integral parts of the equipment, like the one for filling the tires. Not in this case, and so, we learned that this particular task could be performed without the relief valve without ever causing a problem—until it did. A classic circumventing of defenses.

When a step in the process is eliminated, efficiency and convenience increase. That can be a hard choice to combat when removing said step doesn't *appear* to have any adverse effects. You could call it "team norms" or "unsuccessful adaptation." Regardless of the terminology, the more times the task is successfully completed without that step, the harder and more unlikely it is that the circumvention of the defense will be detected and corrected. This is another example that went undetected for years.

THE DYNAMICS OF PRACTICAL DRIFT

The normative force of the factual and omitted safety precautions often leave no trace or need for rework. Modern, integrated IT systems can help to make sure that prescribed tools and materials are indeed used, but they cannot fully preclude shortcuts. The dynamics here are related to "practical drift," a term coined by Scott Snook in 2000 when he described the causes for shooting down two helicopters in friendly fire over Iraq in 1994. Everybody involved moved slowly away from written procedures to optimize their work locally. These local adaptations have been disconnected from each other since teams had only so many interactions with each other, and at one point they didn't fit together anymore. Practical drift has also been included in International Civil Aviation Organization (ICAO) document 9859, which describes the design of the Safety Management System (SMS).

One of the examples Snook provides is the abbreviated language military pilots and other front-line operators often use. In Iraq in 1994, when the F-15 pilots patrolling the no-fly zone were approaching two friendly Black Hawk helicopters, they mistook them for Iraqi "Hind" types of Soviet production and thus enemy aircrafts. The silhouettes of those two models look similar to an extent. Expectation bias and confirmation bias (see chapter 4) made them assume that these helicopters must be enemy ones, but they were not sure. The lead pilot asked his wingman for confirmation after the visual identification as "Hind." The lead said, "ID Hind, Tally two, lead trail, TIGER 02 [the wingman's callsign], confirm Hinds," and the wingman responded, "Stand by," and after closing in, "Tally two," which the lead took as confirmation that they were dealing with two Hinds. The wingman later said that he couldn't and never actually did confirm that these were *Hinds*.

In his scenario, "Tally two" can also mean, "It's your call." If the wingman wasn't sure either, he might have just passed the ball back to the lead pilot, assuming no responsibility. Sometimes people avoid a clear answer by stating facts, stating the obvious, instead of admitting that they don't know, or they are not sure, or they don't want to take responsibility. This behavior is, of course, more likely when talking to your boss, as you subconsciously might assume that whatever you say could be held against you. So, you stay with the facts and leave the interpretation, *any* interpretation, to your boss. An invited misunderstanding. An error trap waiting in the wings.

Abbreviated language can easily conceal that the sender and the receiver don't mean the same thing. "Have you installed the safety pins?" A yes, can mean "Yes, I have installed the safety pins on the main gear, but I don't know for the nose gear." Abbreviated language is just one way where teams optimize their daily work. Practical drift includes the whole spectrum of people, communication, techniques, tools, materials, and everything that is relevant to get the job done, and each and all of those components may move away from written procedure over time. In our case, it was even worse: from day one, the written procedure for torquing the bolts had never been applied.

TAKEAWAY NO. 10: Even more dangerous than practical drift are adaptations which have never been applied correctly.

Another error trap lies in the interface of support and production, which becomes apparent when the official method cannot be followed for one reason or another. In the above cases, it was a fault in the manual or a missing tool. Other cases involve material not available or not clearly interchangeable, alternate equipment, manuals applicable or not. In those cases, the responsibility typically rests with the mechanic who certifies to use "approved data." But it is relatively easy for the mechanic to assume that the support organization has cleared a deviation. That's an error trap. If the tool keeper or the person on the AOG desk offers you a "work-around," you can't take that at face value. It is a bit like the NTO in the engine torquing case. It may look good, but it remains your personal responsibility to validate it, if you intend to certify later to have used approved data.

And, finally, this is not at all restricted to the mechanics at the "sharp end" of the organization. Management decisions can likewise be based on prior mistakes. All too often, templates are used with unverified content. This, for example, can become a problem with contracts, especially with seldomly activated and tested liability clauses. These clauses stipulate who pays the bill if some mishap occurs. Normally they won't be used. You always try to find reasonable solutions in a small world like aviation. Normally you don't meet your customer or supplier in court. Just because nothing bad happened before, doesn't mean it can't or won't happen.

We lessen our mental load by delegating some responsibility for our decisions to support personnel or some previously used template. We avoid asking awkward questions. The inspector who flagged the torquing issue is a lesson for all of us.

CHAPTER 6

Lack of Attention to Detail

*Work with pride, keeping safety
and quality always in mind.*

—JESUS N. "JOEY" VIROLA, HEAD OF PRODUCTION

MANILA, 2008. The flight captain who brought the aircraft over from Hong Kong, I'll call him Capt. Keller, was plainly exhausted and just wanted to go home. On his way to Frankfurt, his flight had to return to its departure airport twice, as the landing gear wouldn't retract after takeoff. Same passengers, same fuel dumping procedure. The line maintenance crew couldn't do anything anymore, so it was decided to bring the aircraft to Manila. Our job was to find the fault, but we couldn't. The landing gear was cycled on jacks several times. Everything was checked as normal. This aircraft had had a normal flight just prior to this landing gear problem. There are only so many things, which could break in normal operations, it shouldn't have been so hard to find. Adjustments fine, sensors clean and working, lubrication fine, not too much grease, all electrical signals present, all pins in all connectors tight, all connections tight, all involved computers replaced, and all operational checks OK, zero fault messages. Nothing.

It took three days to find it: a missing bush (23) in the bogie beam assembly. We could only guess that on a previous maintenance stop, the landing gear was partially disassembled. And with all the grease and dirt, this bush was apparently overlooked after removal and possibly discarded. Once reassembled, you can no longer see the difference. All the tests and even the first flight were uneventful. But then, under minimally different loads in flight, the bogie beam of the landing gear would move slightly differently, enough to convince the aircraft that something was wrong, and the landing gear should not be retracted.

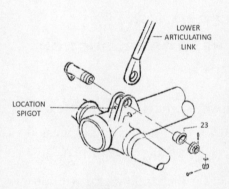

A330/340 landing gear bogie beam. The bush (23) can be overlooked at installation.

A part missing where it can't get lost in flight is the last thing you would expect in line maintenance. When there are problems with landing gear retraction, it is typically linked to adjustment, the sensors, or contamination. But not a substantial part missing, especially not after at least one successful flight. But in the bigger picture, errors of omissions are far more common in aircraft maintenance than errors of commission that involve improper installations or wrong parts. Omissions account for more than half of all maintenance errors, according to Professor Reason: Omissions of parts to install or omission of steps to conduct.

Common parts missed to be installed are O-rings (the L-1011 accident in 1983) or other gaskets, fasteners, caps, and whole panels. Common steps missed are the removal of foreign object debris (FODs), safety devices and tools, proper lubrication, reactivation, testing, documentation, especially the final installation and securing of panels, connections, or other parts when they are just temporarily and loosely put in their place before. Why is that? Why is human work so prone to omissions?

AVOIDING OMISSIONS: IS IT POSSIBLE?

Professor Reason identifies four features of tasks that are vulnerable to omission, and these features explain in turn why humans are so prone to them: Mental Load, Attention, Sequence, and Cues.

Mental Load: We Just Have Too Many Things to Keep in Mind at the Same Time

Mental load is a common human limitation. We can only hold a handful of things in our short-term memory at the same time. If it is a repetitive task, we may use a checklist. However, unofficial checklists bear their own risks. They may be inaccurate or become outdated. Best practices suggest sticking to the official, controlled guidance. One mechanic shared with me a technique he picked up from the Japanese: Use your finger or a pen while you read to help maintain your focus by guiding your eyes to make sure you don't miss anything. Furthermore, we can use independent inspections. In many cases they are mandatory, an inspector's sign-off, but in other cases they can also help us to avoid omissions. And if a colleague is not available to inspect our work, we can also inspect it ourselves by taking time to pause after the task is completed and then return to recheck that we completed all the necessary steps correctly.

If we have too many things on our minds, things might slip off, randomly. When planning a day tour on my bicycle, there were always many things to think of, from charging all the gadgets in advance to the items I needed at the endpoint. I used to forget one thing this time and another thing the next time, until I gave in and put up a checklist with some forty items. That worked. Until I thought I didn't need it anymore, and I had to go back because I had forgotten my water bottles. In the office, it happens quite frequently that we forget to call somebody back, omit a word in a sentence, or miss an appointment which was not in our calendar. These random mistakes may cause delay, inconvenience, or embarrassment but their consequences are often small. The same is true on the aircraft. If we forget to carry a tool with us when we go to a remote position, we might have to go back and lose some time. If we forget to order the gasket with the pump, we might have a delay. If we have ordered the gasket but forget to install it with the pump, chances are that we later see the part still there. Typically, a step downstream will remind us that something is missing. But this is not always the case. That's why we look at particular scenarios below.

Attention: Some Steps Don't Grab Our Attention

Some steps are simply less visible than others, making changes in the manuals easy to overlook. Expectation bias adds to the challenge by letting our eyes see what we expect to see. And in most instances, the old procedure worked fine, so we don't receive feedback even if we haven't followed the new procedure. Visibility also depends heavily on good housekeeping. In a messy workplace, things get lost easily. Housekeeping can, indeed, help or hinder a lot to avoid omission mistakes. Japanese manufacturing companies have pioneered "Lean" production systems, which all start with good housekeeping. Compare production sites or aircraft maintenance hangars for that matter from thirty years ago anywhere in the world with today and the difference is obvious. We have all learned to keep our workplaces in much better order. But look closer and you can still see existing clutter and improvisation, and every little piece of it can distract us from seeing the less visible steps and parts.

And finally, redundant steps, like many safety precautions, also fall into a less visible category. For example, why should we tag a pulled circuit breaker in addition to the safety collar? Here's one example:

Due to a last-minute fault of the forward outflow valve on a B777, the valve was deactivated at 7 percent open for the next flight (per Minimum Equipment List) and its circuit breaker (CB) was pulled and secured with a white collar. While being towed to the terminal, the mechanic on board, who didn't know about this item, saw the white collar on the CB and the status message. He thought it was left over, so he removed it and pushed the CB in. He didn't know it was intended for flight, because he was used to the Airbus fleet rule of red collars for flight and white collars for maintenance.

The reactivation of the valve didn't change the status message on the flightdeck display, but it did trigger the valve to fully open. At this stage, the mechanic realized that the reset didn't help and pulled and secured the CB again. The valve stayed fully open, and the flight took off. The cabin couldn't pressurize, and the plane had to turn back. The red collar/white collar confusion was subsequently addressed in the procedures and standardized between the Airbus and Boeing fleet. But this could have been prevented by strictly following guidance to tag all CBs in maintenance, so that nobody would mistake a non-tagged but collared CB for a temporary maintenance deactivation.

Another example of overlooking a less visible item occurred when airframe vibration forced one of our customer's aircraft out of service.

The subsequent inspection revealed that two of the seven rudder hinge bearings had excessive play and had to be changed. This is not unusual work for the structures team. After removal of the rudder hinge bolt (1), the hinge fitting (3) is separated from the hinge arm (2) by removing the Hi-Lok rivets and a new hinge fitting is installed.

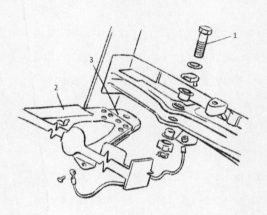

It is January and since last November, Structural Repair Manual (SRM) Rev. 135 was effective for this aircraft, but it had introduced an important change: before installation of the new hinge fitting, the boreholes had to undergo NDT inspection. This was missed and the reinstallation proceeded without NDT inspection.

A320 rudder hinge assembly. An NDT inspection has to be performed before reinstallation.

Later that January, the change was brought to our attention by a similar finding in one of our sister companies. The new NDT inspection could belatedly be done, without findings, after ninety-two flights in February. The mechanics claimed they were still seeing Rev. 134 in the system, without NDT, which sounded implausible but couldn't be ruled out in the investigation either.

Though the consequences here were mild, it showed us once again, how easily an important change can be missed, especially if it involves navigating between different manuals, here the AMM and the SRM, and different configurations, pre or post certain modifications.

Sequence: Some Positions Are Prone to Be Overlooked

In a sequence of steps, the ones in the middle and at the end are more easily omitted than the steps at the beginning. The last task in a sequence is most in danger of being forgotten. Whether we intend to or not, our mind moves on to the next task before the last one is truly finished. We would forget our card at the ATM every second time if it wasn't dispersed before the money. Taking the money is the cue that the transaction is finished successfully. Those cues also happen in normal work. The fault gone, the test passed, the inspector's sign-off done—all signals to move on and possibly to forget the last step: securing the connection, installing the panel, and finishing the paperwork. For this reason, all those "close-up" tasks rank high on the list of omissions.

Close-up omissions are, unfortunately, common. And they all are embarrassing and often high-risk. If the reinstallation of parts or panels has been forgotten, chances are the aircraft will be grounded wherever that is discovered. If they are not securely tightened, they might get lost in flight (as PDA, "Part Departing Aircraft"). If test equipment or tools are left installed on the aircraft, system functions might be disturbed or lost. When you leak-check the fire-extinguishing piping on an A330, you disconnect the pipes and install four blanking plugs, but you, of course, must remove them again and connect the fire extinguisher bottle before sending the aircraft back to service. If you forget that, and believe me, it can happen, you can have this engine and more so, this wing unprotected in case of engine fire, and chances are that the fault would only be seen at the next A-check, some six weeks later. The customer may or may not send you back your blanking plugs with love.

Cues: Some Steps Are Prompted By Others, While Some Are More Isolated

If a step in the sequence is not naturally prompted by a preceding one or by an obvious state of the aircraft or part, it may be forgotten. If you see a hole or a gap, it prompts you to install the missing part. If you see a remove-before-flight flag, it prompts you to remove the item. But take the pitot probe covers, which must be removed before powering on a parked aircraft. There is not much of a cue to remind you of that when you sit in the cockpit. It is the same with the removal of the tail stay when lowering the aircraft from jacks or closing the thrust reversers before extending the slats. All these steps, that are not so obvious and lack prompts to remind you to check for these items, are easier to overlook.

Let's look at a specific example: safety pins. Often ground equipment for transporting big aircraft parts uses safety pins to prevent parts of the structure from moving and to keep the component firmly in place. It is common that those safety pins are secured by a lanyard, so that they don't get lost when not in use. But sometimes those safety pins have a retaining clip on their own to keep them in place once installed in their intended position. And if this clip gets lost, there is not much (other than the hole in the pin) to prompt you to install the clip. An A320 main landing gear leg, installed on a transport trolley, fell and hit a mechanic's head (the mechanic recovered, the aircraft part did not). It was caused by a missing retaining clip on one of the safety pins. The mechanic, who installed the safety pin, wasn't aware of the retaining clip as it was out of sight or missing. Afterward, we started to also secure the clips with lanyards.

Omission Error Traps Are Everywhere

Omissions are a very big part of a lack of attention to detail. But there are more. Let's look at mistakes out of convenience, for lack of a better word: taking what is at hand without much consideration. Cross-connections are common and potentially a very critical mistake. Cross-connecting mechanical elements or electrical wires can lead to inverted left/right or up/down input to flight controls, which may cause loss of control and even loss of lives as it did on May 30, 1995, when a pilot was killed when his F-15 crashed during takeoff on a routine training mission.[9]

Fortunately, not all cross-connection omissions incur such drastic consequences. For example, when the sense inputs to the A330 anti-ice valve are cross-connected, items 3 and 4 in the figure below. In this case, the connectors are the same and the lines are flexible enough to cross them. The manual warns:

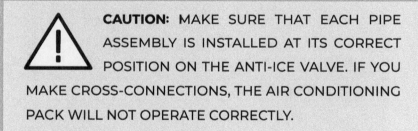

CAUTION: MAKE SURE THAT EACH PIPE ASSEMBLY IS INSTALLED AT ITS CORRECT POSITION ON THE ANTI-ICE VALVE. IF YOU MAKE CROSS-CONNECTIONS, THE AIR CONDITIONING PACK WILL NOT OPERATE CORRECTLY.

If it happens, the temperature in the cabin will not be regulated as expected, and it might be hard to troubleshoot. It is a problematic design like we have seen before, but at least the manual provides

---

9   John Diamond, "Two maintenance NCOs face courts-martial in F-15 crash," AP News, October 14, 1995, https://apnews.com/article/8e0bc96b1ae9bc4b50e4e62544 629a92.

proper guidance, and this being the only CAUTION note in the Installation of the Anti-Ice Valve task should grab adequate attention.

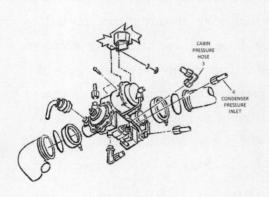

CABIN
PRESSURE
HOSE
3

4
CONDENSER
PRESSURE
INLET

Taking what is at hand, not what is prescribed, or like in this case completing the installation as it seems to make sense, is a permanent temptation for every mechanic. As many fluids and lubricants look alike and they may (wrongly) be considered less important than the hardware, mistakes of choosing the wrong one,

A330 anti-ice valve. The cabin pressure hose (3) and the condenser pressure inlet (4) can be easily cross-connected.

the one which is at hand, can have unexpected and grave consequences, too.

TAKEAWAY NO. 11: Attention to detail should be our second nature in aircraft maintenance.

Once again, *competence, awareness, compliance,* and *teamwork* help to prevent errors of omission and cross-connection. Especially the *awareness,* considering how easily and naturally we forget a step, or we do something automatically without checking whether it is correct. I encourage all of us to embrace the defenses, like independent inspections and checklists, that are available to us.

CHAPTER 7

Missed Handover

Misconducts damage the profession. How can we go about changing big things if we cannot follow simple rules?

**—ALEX E. PEÑASALES,
SECTION MANAGER, PRODUCTION QUALITY**

MANILA, 2019. The A380 movable flap track fairings (MFTF) are big components and have elaborate inner workings. In one of the C-checks, a service bulletin had to be performed for a special detailed inspection (SDI) of the kinematic elements of MFTF #2 and #5. The parts were removed, tagged, and routed to the NDT shop. When they were received back, the similar-looking parts were installed again following the location marked on their removal tag. Unfortunately, the operating rod (1), one of three parts, was mixed up between location #2 and #5. It could be fitted indeed, but the parts are not equal. One is shorter when extended. When the flaps were later cycled with the rods in wrong positions, the left-hand inboard flap and the MFTF #2 were damaged. The mechanics had not checked the right part number for each location according to the IPC. Whether the removal tags were mixed up in the NDT shop or they checked at all

(all four rods for two locations on both sides looked the same), couldn't be established.

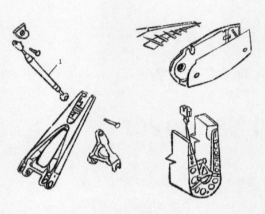

A380 movable flap track fairing. The operating rod (1) can be installed at different positions.

This is our first example of a secondary error trap. The defense for mixing up parts was not just weak but somehow clumsy and made matters worse. The mechanics relied on the temporary removal tag which was apparently swapped in the process of NDT. Without the tag, which is meant to guarantee proper parts segregation, they could have checked the part number on the rod and avoided the mishap altogether.

Scenarios like this one can happen with any handover of information from one shift to the other. On another occasion it also involved the flap track fairings. As one shift couldn't complete the removal of all flap track fairings, they handed it over—unfortunately, without all the details. And this is so often the case.

Their task was "simply" to remove the flap track fairings of this B777 aircraft as a preparation for its C-check. The aircraft was still parked outside of the hangar which is normal in the first one or two days for incoming checks, defueling, or simply because the hangar is still occupied. If you remove (or install) big parts outside, your attention can be easily consumed by the difficulties of access (platforms, lifting devices, cranes), the weather, and lighting. From the incident report:

- The work done report in the logbook listed the following:

  - Flap track fixed fairings no. 1, 2, 3, and 6 removed.

  - Flap track movable fairings no. 3 and 6 removed.

  - Flap track fixed fairings no. 7 and 8 access panels removed.

- The workload handed over to the next shift were:

  - Continuation of flap fairings removal.

  - Tail and wings panel removal.

  - THS lubrication.

- The weather condition said occasionally raining at that time.

What the report didn't mention was that a cross tie was left and tied in the flap track of LH outboard flap support fairing no. 6 during the lowering of the fairings. The mechanics anticipated that the morning shift would continue the fairing removal and easily see it.

But the next shift didn't continue the removals outside but rather repositioned the aircraft inside the hangar. For towing they had to retract the flaps. And that partially disassembled cross tie got in the way and damaged the fairing. From the report:

"Flap Support Fairing no. 2 aft lower surface was struck by the loose support link during the flap retraction. Consequently, Frame nos. 1 and 3 were damaged by the movement of loose parts inside the fairing. Cross tie/ bracket, considered as FOD, was left and sandwiched between flap track no. 6 mechanisms during retraction leading to its deformation."

A little change in circumstances created a blind spot, an error trap. This was the first technical incident in this production line in three years. As technical incidents, we count all accidental damages over a certain threshold, all mandatory occurrence reports, and all in-service problems after redelivery of the aircraft. Where was the supervisor? Well, he was recently promoted to lead another flight control team in another line but was still assigned from time to time in his old bay in parallel. As it happened, after he assigned the mechanics to their jobs for the removal of the fairings, he was called to lead the towing of the aircraft preceding the B777 on its day 1.

That aircraft was delayed and thus overlapping with the next project, because the customer couldn't deliver some parts on time. But now it was ready for release to service. Later that evening, he was called to supervise an A340 flaps zeroing in his new job. After all these interruptions, he returned to his removal team late. It didn't cross his mind to check with them for any items which could impede flaps retraction. He also didn't order the removal of the scaffolding as he thought the work would continue outside in the morning. But, yes, the early shift team leader had mentioned to him ten hours ago that the aircraft would be towed the next day. These are the messy details of normal work.

Information can get lost despite our specific means to capture them: removal tags, turn-over logbooks, and checklists. And sometimes these defenses create new problems as people rely on secondary data without looking anymore at the original documents and the actual, physical state of the aircraft, but they could be outdated, incomplete, or outright wrong as well. So, we must extend our repertoire from "respect the defenses" to this:

TAKEAWAY NO. 12: Defense measures themselves could become error traps.

The defenses in place to avoid the loss of detail in the handover of information, can unfortunately, create errors themselves *if* people rely on the information they see, and that information is faulty. Again, "It ain't what you don't know that gets you into trouble. It's what you know for sure that just ain't so."

It could also be the other way round. In one instance the structures team had to replace the fire seals on all four thrust reverser halves of an A320. The work was completed, and the aircraft was released to service with five fasteners not installed yet but with their dummy fasteners, known as clecos, still in place, see the picture below.

Delivering work that is so obviously incomplete, is of utmost embarrassment for any maintenance team. So, what happened? The mechanic, who had done the work up to the installation of the last five of fourteen fasteners on this thrust reverser, had been reassigned to another project. He documented his work carefully on the two non-routine cards (one for each engine side, LH and RH) for the four thrust

Remaining "clecos" in 12 o'clock position. The modification of the thrust reverser was incomplete still.

reverser halves. All four units were removed, now sitting on the hangar floor. He couldn't distinguish which of the four belonged to which engine side and, thus, to which non-routine card, so he documented his work for the two already completed on both cards, and he put "completion of fastener installation" to be done in the turn-over logbook.

The next shift then saw the completed non-routine cards and thought the logbook entry now obsolete. The final mistake happened on the part of the inspector, who didn't use a ladder for his final inspection and the cleco fasteners were hard to see in the twelve o'clock position. Also here, the overlap of a delayed project with a new one added stress and distraction for the mechanics. And a further defense didn't work as this team used "their own" clecos, not the controlled ones from the tool room, otherwise the unreturned tools would have prevented the premature release. This last finding triggered a company-wide search for unauthorized tools, which I think all MRO companies have to do, annoyingly regularly. It seems to be too tempting to circumvent the tool-control bureaucracy. Another example of what Snook called "practical drift."

So, here the logbook was correct, and the documentation in the task cards was wrong. A word from Vanessa Shawver, an instructor from Tony Kern's team, comes to mind: "If there is doubt, there is no doubt." If something is not feeling right, it is probably not right. These three examples all look like particular error traps for the mechanics as the normal defenses were disturbed: in the first case, the removal tags were interchanged; in the second, the handover information was incomplete; and in the third, the handover information was contradicting the primary documentation and thus dismissed. In none of these cases were the mechanics charged with negligence and subjected to disciplinary action, but I refuse to accept incidents like these as unpreventable.

Let's check back with our four tools: *Competence* should make you suspicious of any handover. There are always ways to double-check the handover information with the real state of the hardware.

Awareness should shift one gear up if you cannot verify the handover information. In the first case, all four rods coming back from the shop looked the same but their installation locations were not, a strong cue to check the part numbers. In the second case, the work was interrupted at an unusual point and turned over to another team in the morning shift without any physical contact, a strong cue to be precise. And, finally, in the last case there was already a mismatch in the documentation which should be a powerful cue in itself.

If all that failed, meticulous *compliance* would have still prevented the errors. Checking the part numbers with the IPC before installation would have prevented the mix-up. Checking the flaps system before retraction, especially after seeing that the last shift obviously did not expect towing as the next step but rather the continuation of removals. Inspecting the whole area, including looking for a ladder when appropriate. And, finally, *teamwork*. When it comes to handovers, we should make extra efforts to provide extra details and warnings to the next shift, and, if we are the next shift, we should not take everything automatically at face value. "Trust is good, control is better" (often ascribed to Lenin).

That reconfirms:

TAKEAWAY NO. 13: If there is doubt, there is no doubt.

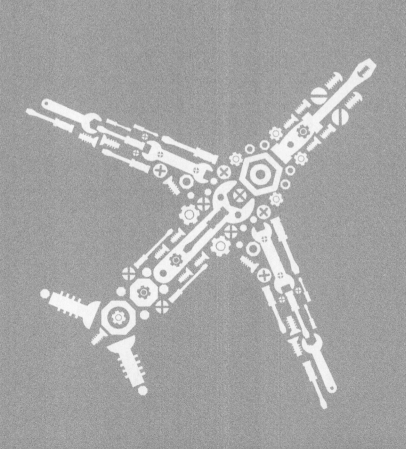

CHAPTER 8

Unsuitable Procedures

In a complex system, because mindless compliance doesn't work, you have to be mindfully non-compliant.

**—GEORGE DOUROS,
SAFETY, CURRICULUM DEVELOPMENT**

BUDAPEST, 2012. The towing incident. One of our best mechanics sat in the captain's seat. His crew was fighting to get this A320 back into service. The customer was constantly on the phone with him. The last step was the engine ground run. As we use a run-up spot outside of our premises in Budapest, the aircraft must be towed out to the taxiway. On push-back from the hangar, wing walkers guide the movement until the aircraft reaches the centerline parallel to the hangar. Then all the marshallers and wing walkers get into the bus which follows the aircraft on its way to the designated area, as the way forward is clear, and the tow truck driver controls the movement alone. At least that's what was expected.

An aircraft was parked in front of the western half of the hangar, appearing to be out of the way. The tow truck driver had no other guidance than to stay on the towing line. The towing crew leader in the captain's seat had no role other than to brake in an emergency, and while en route, he was on the phone with the customer. Given the circumstances, the inevitable happened: the left wing touched one wing of the parked aircraft. The collision damaged the leading edge of one aircraft's wing and the trailing edge of the other one's wing. Massive repair cost and time, as usual for towing incidents.

Whose fault was that? The procedures didn't foresee this specific scenario, but they oblige the towing crew leader to always make sure that he has enough clearance, otherwise to stop and get his wing walkers out. Additionally, he should have known better than to let the customer distract him. Because of the frequency of towing movements and the possible (and real) impact of incidents, towing procedures, in particular, can easily involve clumsy defenses. Sometimes procedures are so overloaded that nobody knows exactly how many people are needed at a minimum and exactly how interactions should occur in varying conditions.

This was a landmark case for me personally. Up to that day, I believed people should simply be more careful in doing their work. If they followed procedures, accidents wouldn't happen. And if procedures aren't followed, a punishment that teaches them a lesson once and for all would be required. But this case gave me pause because it involved one of our best mechanics. OK, even worse then, because the best mechanic really should know better and therefore he deserves an even harsher treatment or so would be my natural train of thought. If there is one positive aspect from this event, it is that it initiated my interest in "human error." My belief that everyone should always follow all procedures or face punishment just didn't feel right for this scenario and so, began my questioning to find out why.

For scientists, like Sidney Dekker, James Reason, and Nick McDonald, it is already common sense that the best mechanics are not by chance involved in the worst accidents. They are taking responsibility in difficult, non-standard situations. The best mechanics often operate at the edge of what is possible. They interpret the rules and procedures in unforeseen circumstances and, if things go well, we appreciate their can-do attitude. But if things end badly, we see judgment error and too aggressive risk-taking. In hindsight, things look so obvious, but at the time of the action, time pressure, limited information, and bad luck may have cornered our team member in a way that made an error quite understandable.

Highly skilled mechanics, and other front-line operators for that matter, lose their livelihood because they have pushed it too far. The recognition of this fact has prompted me to write this book. Not to tell our team members that they simply must try harder, but to remind all of us that our world is full of error traps. It is a hostile environment that requires extra effort to survive. And on the part of the leaders, if you praise your best mechanics nine out of ten times for pulling it off under the most impossible of circumstances, then be careful to judge them on the tenth occasion for violating procedures and thus willingly putting safety at risk.

A friend told me some time ago they had too much work, but they had already committed to a landing gear change for a very important customer. There was zero hangar space and no hope of obtaining workable space. So, they directed the aircraft to a secondary airport that didn't even have a hangar. They performed an intense risk assessment, observed the weather as best as they could, considered emergency measures, and jacked up the aircraft outside. They did the full landing gear change completely outside in the open, successfully, but illegally. Customer was happy. They pulled it off. When I first heard it—and this is why I share this delicate episode here—I, too, thought, "Wow, they made it happen." My first reflex was admiration. And that's how your best mechanics learn to test the limits.

I guess I am not alone with admiring successful operations at the edge of the safe and legal envelope. But it is wrong. We cannot justify transgressions like that when they are successful, and we shouldn't admire them either. When we do, we send the wrong signal. And as I thought it was so easy to end up here, I thought I should write it down.

THE EVOLUTION OF UNSUITABLE PROCEDURES

Let's return to the towing incident. Towing is like a micro-cosmos of accident-prone aviation maintenance. The slightest contact has incredibly expensive consequences. Towing is a low-skill, slow-motion activity (except for the driver, but he has to rely on the team), which invites people to do it on autopilot, not really in the moment, not taking it seriously, and the procedures are often over-specified by people who have never done the work themselves. If it goes wrong, penalties are harsh as the damage is big and people are expected to be more careful. Arguably, how your team tows an aircraft (and how you process towing incidents), is an indicator of how safe your operations are.

Unsuitable procedures under the control of the MRO would signify what Professor McDonald would call "sloppy management." But what about Original Equipment Manufacturer (OEM)-prescribed procedures that are less than optimal? The official manuals? McDonald and also Helmreich conducted famous studies that showed how often mechanics or pilots, doctors, and nuclear power plant operators for that matter, deviate from the written guidance and how seldomly that leads to problems. And when procedures prove indeed unsuitable, the change process is often slow and tedious. As a result, the operators are accustomed to functioning with suboptimal procedures.

The mandatory SMS, based on ICAO document 9859, leads us back to Snook's "practical drift": over time people seek more efficient ways to perform a task, and deviations from the procedures become "norms." The SMS shall ensure that audits detect these deviations and trigger corrections, so that the work returns to the letter of the guidance, or the guidance is adjusted. In practice, this is extremely difficult. For scientists like Hollnagel or Dekker, this tension often causes unjust treatment of operators accused of human error and prevents very safe systems like aviation to learn, innovate, and further improve safety.

Let's put unsuitable procedures firmly on our radar of potential error traps. As users, we must be aware that any adaptation must stay within safe and legal boundaries and that violations of the airworthiness of the aircraft have not been incurred. That implies that you have done a proper risk assessment and that you are entitled to take this risk. As managers, we must take our responsibility seriously and acknowledge how significantly unsuitable procedures create error traps and push our teams toward possible violations. We must also acknowledge the level of effort we must exert to tackle unsuitable procedures. To recall Professor Reason's quote from chapter 2: "One of the defining characteristics of a safe organization is that it works hard to find and eliminate its error traps."

VIOLATIONS: THE CONSCIOUS AND THE CASUAL

To borrow again from Reason, violations can be put into five categories: "routine violations," "optimizing violations" (sometimes called "thrill-seeking"), "necessary violations" (sometimes called "situational"), "exceptional violations," and "unintentional violations."

- Routine violations are associated with practical drift and team norms. They can be considered bad without exception (either the rule or the deviation is bad).

- Optimizing violations are self-serving deviations to make the work more convenient or exciting.

- Necessary violations are seen as inevitable to avoid work stoppage.

- Exceptional violations are linked to rare and unusual circumstances related to emergencies and don't play a role in daily life.

- Unintentional violations follow lack of knowledge and awareness of the rules.

While optimizing, unintentional, and necessary violations may be understandable, they are hardly defendable. That does not really leave space to justify as many as thirty-five out of hundred tasks deviant from the procedure as the authors Helmreich and McDonald suggest, and maybe it doesn't even justify one single case. But is that realistic? Is it desirable to eliminate violations from our world entirely? This is a big question not only of safety management, but of life, and I don't dare to answer that here in passing. But let me try this: Instead of looking at five categories, let's build just two: conscious violations and casual violations. Both can land us in prison to be sure.

Conscious violations could be considered a small subgroup of situational violations, which constitute conscious decisions to deviate from a rule and to consciously accept a certain risk (at the least, the risk of sanction plus a possible risk of adverse outcome) to attain certain benefits that outweigh the risk. These conscious deviations can, per definition, only happen after a conscious risk assessment. I am not saying that these are good or bad decisions. All other violations, which are not the result of a risk assessment and a rational decision process, we'll label as casual violations and consider them undesirable per se. Driving as fast as the other drivers without looking at the speed limit could lead to a casual violation. Limiting the overspeed to +10 percent, on the other hand, would constitute a conscious violation.

A captain who brought his aircraft to our MRO had to take one of two entries from the taxiway. We were waiting at one of these, but the tower directed him to the other, which was already congested on our side. He asked again, the tower declined again, and then he decided to take the one he and we wanted. A clear violation. He did a risk assessment, and he made a conscious decision. Had he followed the tower directive, we would have ended up with hours of towing aircraft back and forth to resolve it. He told me later that we cannot do that too often. Indeed. He took the risk of sanction (plus minimal additional outcome risk) for the benefit of saving us hours of decongesting the situation. I am not saying that this was necessarily a good decision. But it was a conscious decision.

We find conscious violations among situational violations. Reason describes them as resulting from a mismatch between work situations and procedures. It's likely that most situational violations occur without a risk assessment. This is hard to validate in practice and a risk assessment can of course be informal and brief. It seems implausible, however, if as many as one-third of all tasks feature deviations from the guidance, that more than a few of them are the product of some sort of elaborate risk assessment which takes consequences in outcomes and sanctions into account. Most deviations in practice appear to stem from the casual sort, conscious ones are rather the exemption.

If people are involved in mishaps that we assign the "human error" label, and if part of it was a departure from procedures, we must be careful to assume they consciously decided to deviate, thus made the wrong choice. In practice, most decision-making is fast and based on experience, heuristics, and common patterns, rather than slow, rational, elaborate, and comprehensive. If you are caught in a speed trap, try to remember when you decided to speed. You didn't. It happened. It's different with a red light. You probably never run a red light or cross a "red line" without a conscious decision.

In modern safety management, much has been done to understand, re-label, downplay, and even justify violations. Complex, high-risk socio-technical systems are seen to have only a narrow band to operate in both efficiently and safely, and this operating point can easily migrate under production pressure and individual optimization toward the edge of the safety margin and beyond. When that drift happens, small variations which normally don't matter much, including violations, may push the system into failure. In such a scenario, a violation is just a minor factor, not more important than perhaps the weather. Migration happens, for example, when we grow accustomed to working with unsuitable procedures, the starting point of this chapter.

At this point, though, we must be careful to prematurely "decriminalize" violations. All too easily we throw out the baby with the bathwater. If we accept for a moment that we see casual and conscious violations, fast and slow violations, if you will, we can say that casual violations are undesirable as their consequences are not taken into account as part of a fast unconscious decision-making in which the transgression doesn't play a role. It's more difficult for conscious deviations. We take the consequences into account and bear the risk of sanction. Still the benefits outweigh these risks. These situations strip us of our most potent excuse that we didn't fully understand. We *did* understand the "red line" but decided to cross it. A bit like pleading "not guilty" and battling it out.

At the gate, a captain on a flight from Seoul back to home base in Europe received an unfortunate call informing him that maintenance missed an update of his navigational database, and it expired yesterday. What shall he do? Offload all passengers again to the hotel and cancel the flight? How big is the risk that a waypoint just changed yesterday and that it would have any influence on his flight that he couldn't handle? On the other hand, it is illegal to fly knowingly with an expired navigational database, and everybody he could call would tell him exactly that. What did he do? He made the conscious decision to save the company and its customers a lot of unnecessary hassle and he took off from Seoul as scheduled.

It so happened that there were thunderstorms in the area after takeoff, and as he tried to avoid them, he slightly touched North Korean airspace. This airspace violation triggered an official investigation in which it was established that he should have never taken off in the first place, and the captain might find himself fighting for his license before long. This is an extreme (and hypothetical) example. But still, if you violate consciously, you should have a plausible risk assessment at hand if you have to defend yourself afterward in court. Thankfully, the airline which provided this example has meanwhile adjusted their procedures to legalize such a decision.

But we find both in this basket of conscious deviations. The judgment errors (or the coincidences of life, which sometimes expose seemingly good decisions) and the weaknesses of the system which we have no other sensible way to deal with. In this basket, we must look at the scenarios case by case. We must reflect on our red lines which we don't cross casually. Confronted with unsuitable procedures, we must decide to muddle through or battle it out, which might end up with work stoppage or, in the extreme, a criminal offense. I dare to say almost every day, we all must make decisions like this, and we must make them consciously, whether jail is a possibility or not.

TAKEAWAY NO. 14: Casual violations are bad. Conscious violations are part of life (but may be bad nevertheless).

CHAPTER 9

TL:DR (Too Long: Didn't Read)

MISSED NOTES, MISSED INDEPENDENT INSPECTIONS

Safety is safety regardless of the pressure we feel.

—LUCKY R. OCAMPO, CREW CHIEF

MANILA, 2017. An A330 aircraft fresh out of maintenance flew on its first flight to Haneda (HND), Japan, and was declared AOG there due to high engine vibration. The pilot experienced 5.5 units N1 vibration at takeoff and during climb. Power reduced; the advisory disappeared for the rest of the flight.

Modern aircraft shield pilots from deviations unnecessary for them to know. Only when it reaches a certain threshold (3.3 units in this case) will an advisory prompt them for action or at least consideration, like reducing thrust in this case. The A330's Trent-700 engine is a bit notorious for vibration which originates from the big fan blades of the low-pressure compressor. These blades require elaborate balancing and lubrication care. When we saw the first ones in our maintenance in 2006, we were provided training videos to assure the lubrication was done accurately. Eleven years later, was watching those training videos still standard procedure for all new mechanics? Perhaps not.

In this instance in 2017, a young run-up operator during his first high-power run or "vibration survey," thought 1.6 units were good enough. It was—sort of. He overlooked that the manual required an EPR of 1.54 (1.592 for the day, so 1.54 would be the limit to obey).

> *(4) Slowly increase the tested engine speed to the maximum static Engine Pressure Ratio (EPR) for the day or 1.54 EPR Ref. AMM Task 71-00-00-860-836, whichever is the LOWER.*

The young mechanic said he experienced strong aircraft movement at 1.495 EPR, so he didn't increase further, and 1.6 units vibration is already high and exactly on the troubleshooting level.

> *(d) Vibration survey troubleshooting level. If the LP (N1) vibration from test No. 11 the engine vibration survey is more than this level, then do the troubleshooting:*
>
> *1 LP band 1.6 units (0.5 ins/sec peak velocity) Ref. TSM 77-00-00-810-814 or Ref. TSM 77-00-00-810-815.*
>
> ***NOTE****: The Troubleshooting of engines which have LP (N1) vibration of more than 1 unit (0.3 ins/sec peak velocity) during ground running is recommended to prevent too much vibration in service.*

But the NOTE also says troubleshooting is recommended if N1 vibration is higher than 1 unit. The mechanic didn't read that far.

The documents he had to read for the entire vibration survey are just five pages, two for the run-up itself and three for the limits. This is shorter than for the opening of the FCDs. Yes, there are times when manuals are too long, making them a challenge to fully digest, but not the case for this one. And it was this crew's first vibration survey. And they didn't reach the required 1.54 EPR power level.

It is sometimes tempting to blame the operators for apparently easy mistakes when you have all the information and look at their decisions in hindsight. We should always be very careful with that. But where was our defense here? The inspector was there indeed, but he "trusted" the operators, he didn't challenge that they had ignored the recommendation and he didn't challenge whether 1.495 EPR would be enough, and, after all, they were within the limit. So, this defense didn't work very well. Multiple inspections can indeed constitute clumsy defenses. When all the facts were known, it became apparent that only two blades had been changed and lubricated, but not all, which was an acceptable procedure. The fact that relubrication of all blades at HND fixed the problem, points to the consideration to preemptively lubricate all blades in such scenario, at least when vibration is above 1 unit.

The recounting of that scenario missed some information that will help in understanding, so let's look at it again in detail, and let's look from the following error trap: Too long and/or simply didn't read. In this case, it wasn't that the instructions were too long, but that the mechanic simply chose not to read further because his test returned a "within the limit" result.

This story begins with the replacement of two fan blades on a routine C-check. Modern jet engines produce their thrust mainly with the big fan, or low-pressure compressor, which you see when you stand in front of the engine. The fan is driven by the turbine in the aft part of the engine, which is powered by the combustion of the fuel. The turbine produces thrust itself but most of its power is used to move the big fan like a propeller. This way, most of the airflow doesn't pass through the turbine, but instead bypasses it, which drastically reduces the noise modern engines emit.

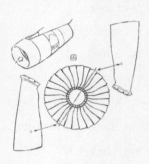

So, if one of the big blades is damaged and must be replaced, which is quite common from impact of small stones, birds, and other foreign objects in the airflow, it depends on the difference of the moment weights of the old and the new blade, whether the opposite blade also must be changed. To avoid vibration, all blades must be carefully trim-balanced, which means small weights are added to account for production tolerances and wear.

Fan blades ("LP compressor blades") of the Trent-700 engine. When replacing fan blades, it may be necessary to do that in pairs. In any case, the moment weights used for balancing manufacturing tolerances have to be taken into account.

This is what happened in our case. The two blades were lubricated, installed and the aircraft subjected to a vibration survey, which is an engine ground run at high power. The other blades were not re-lubricated, but this has been an acceptable procedure as per AMM.

At the run-up, they got a quite high 1.6 units of vibration even at a lower than prescribed power setting. The manual says that 1.6 is the acceptable limit but recommends troubleshooting beyond 1.0—the missed information. The mechanic only focused on the critical level of 3.3, which would trigger a message for the pilot during the flight advising them to reduce power. They just didn't read far enough to acknowledge the note regarding the troubleshooting limit. The sensible action for that troubleshooting would have been relubrication of all fan blades, which was finally done at the outstation in HND. That brought vibration back below 1.0 and cleared the problem.

Now, having looked at this case in a little more detail, it is also noteworthy that the inspector chose to simply trust the mechanics. Scott Snook, who we have met earlier with his friendly fire case study, calls that the "fallacy of social redundancy." The mistaken belief is that more people than needed would create a better result. What often happens in practice is that when more than one person must sign off for a task, responsibility is diffused, and people rely less on themselves and more on each other. Inspectors must be highly alert to the reality that they are responsible for not only catching the ordinary mistakes mechanics make, but also those that are made because of the perceived safety net of the inspector's inspection. This is not confined to the shop floor. In the office, you can find many papers with multiple sig-natories where it is at least questionable whether everybody is reading them before signing. And there is often less expectation from people in meetings when their boss is also present.

One final consideration before we move on. When I was collect-ing the facts for this case from 2017, I couldn't believe that this could happen despite those red flags: the engine type notorious for vibration, the mechanic's first vibration survey, the high vibration reading, the EPR not reached. So, I found the operator from back then, and I talked to him. Did he really overlook the note? He did. It was his first and thus far last incident, and he couldn't explain it either. Now, he is a diligent instructor for the new, young mechanics. Maybe he was intimidated by the tremble of the high-power run, where the aircraft shakes so violently that you can't read from the paper anymore.

Maybe he didn't want to assert himself in a group with a lot of highly experienced colleagues, and when they couldn't set power high enough, they might have experienced some plan-continuation bias, to get over it and finish the test. Maybe he trusted that high vibration after maintenance often calms down by itself after a few cycles. For our topic here, TL:DR, he may have already developed some optimization to read only selectively, which is a required skill. But he definitely learned from his mistake.

OUT OF THE ENGINE AND INTO THE AIRFRAME

Let's get out of the engine and into the airframe, specifically the cabin floor. Our first case deals with the painting of small panels at the entry doors (door sills), which require a special gray anti-slip surface paint. A finding at a sister company triggered an investigation revealing that we had used the wrong paint for some time on a couple of aircraft. Paint that did not comply with the manufacturer's instruction. The paints we used were gray and anti-slip but only approved for exterior use (on the wings) and gray and interior but not anti-slip. Besides its anti-slip property, the prescribed one had different flammability and toxicity specifications for interior use. But this came into effect via a change in the manuals which was overlooked at more than one site.

To identify the correct material part number to withdraw, the painters should resolve the reference they see in the AMM or SRM with the use of the Consumable Material List (CML). This search for any changes must be intentional on the mechanics part, as the manuals do not offer obvious "Look here for change" prompts, and once in the CML, you don't see the restrictions anymore. In this case, restrictions prevent the use of certain paint on internal structures, e.g., doors sills. If you have eight different gray paint products at your disposal, it can be confusing. But no excuse. Yes, the references between manuals can be tedious and changes can happen without a prompt, but they must be read and understood.

What I find striking in this case is that even a relatively miniscule part of the overhaul of the aircraft, like the repainting of the door sills, could result in a product recall, a working party sent around the globe or even an immediate AOG including flight cancellations, and that the painters, and no offense to painters or the workshops here, are just as much exposed to the risk of invalidating the airworthiness of the aircraft as the releasing inspectors at the other end of the pecking order, if you will. Plus: They might have thought (or subconsciously assumed) that the flammability and toxicity of those small amounts of chemicals are nothing compared to, for example, the three tons of random hand luggage and wardrobe which a fully loaded widebody carries. Again: no excuse. The smallest practical impact at the hands of the mechanics least concerned with the airworthiness of the aircraft can have stark consequences if manuals are not followed, and not reading everything poses in a way a more pronounced error trap in the paint shop than in the inspectors' office.

Another example from the interior: The repair of the cabin floor structure is one of the main reasons for a substantial hangar layover, as fluids and moisture penetrate the not-so-easily accessible areas under galleys and lavatories, and the removal and reinstallation of all the cabin items is quite a lot of work. The last step after the repairs is re-protecting the structure with corrosion preventive compound or what Airbus calls a Temporary Protection System (TPS). The AMM offers you a two- (Type 1 and Type 2) and a one-component compound (Type 3). The two methods are interchangeable except on movable components, but that doesn't apply to the cabin floor.

The manual instructs:

1. If TPS Type 1 has been removed, apply the same TPS Type 1 to the surface.

2. If TPS Type 2 has been removed, apply the same TPS Type 2 to the surface.

3. If TPS Type 1 and TPS Type 2 have been applied in an inspection area and are removed: Apply the same TPS Type 1 and Type 2 or, as an alternative, apply the TPS Type 3 to the inspection area.

4. If TPS Type 3 has been removed, apply the same TPS Type 3 to the surface. As an alternative to TPS Type 3, apply TPS Type 1, Type 2 to the inspection area.

While in other areas our customers are very specific as to whether they use, for example, Skydrol or Hyjet, which are both fire-resistant phosphate ester-based hydraulic fluids (but shouldn't be mixed with each other), they typically don't insist on a specific TPS Type 1 and 2 or Type 3. If they do, they issue an additional document to amend the AMM. In our case, the customer directed the use of Type 1 and 2 in an engineering document. This directive wasn't followed by the team. They instead applied TPS Type 3, and the error was not discovered until after the cabin had been completely re-installed. That error resulted in five days of rework which meant five days of layover extension.

When a mistake like that happens which has minor or no technical consequences, we typically ask Airbus for concurrence with a deviant situation, at least up to the next C-check where we could correct it. This wasn't possible here, as it wasn't even a deviation for Airbus. We had followed the AMM but neglected to follow our customer's amendment to the AMM. Again, small things can have big consequences. Small deviations surely can, and small may look like no deviation.

Aircraft or component manuals and all related documents may be hard to read, tedious to navigate, and, most challenging of all, they contain a tremendous amount of frequently changing information. In earlier times, MROs translated all manuals to the local language, and they laid out job cards, also in the local language, handling specific non-routine scenarios. This is not viable in aircraft maintenance where every skilled mechanic must use the original manuals. But the manuals also have become better structured and with the help of modern MRO software, manual references and inventories can be connected by hyperlinks. Software can prevent people from using the wrong procedure, parts, or tools.

TAKEAWAY NO. 15: There is no excuse for not using the current manuals and instructions. All of them.

What about *selective* reading? Our brain is good at focusing on the relevant bits, not only when we read. Only by discriminating the signal from the noise can we cope with what seems to be overwhelming. In our examples, people went too far down that road ("I have what I need, so I don't read more") or they didn't fully appreciate the many ways the manuals could change unexpectedly or be amended by other documents. The biggest threat is that nothing visibly adverse happens and a deviation gets normalized, and when that happens, multiple aircraft can be affected and rework and loss of reputation multiplies.

Finally, "nothing visibly adverse happens" can sometimes take years to prove itself wrong, like in the horrible scenario of Japan

Airlines Flight 123. It remains the deadliest maintenance error to the present day in which "nothing visibly adverse happened" for years. In 1978, flight JA8119 suffered a tailstrike during a landing. This damaged the rear of the aircraft's fuselage, as well as its rear pressure bulkhead. The damage was improperly repaired, resulting in the development of fatigue cracks affecting the rear bulkhead that went unnoticed, or perhaps noticed but not addressed, for seven years before the damaged bulkhead could no longer withstand the pressure changes experienced in flight—*seven years*. Its failure in flight caused the crash that killed more than 500 people.

A crash that everyone in aircraft maintenance should remember, not only for the tragic loss of life but also to serve as a keen reminder that our mistakes, if not recognized and addressed, can sleep for years, slowly and invisibly eroding the safety of the aircraft allowing tragedy to strike, in this case, so much later.

CHAPTER 10

Miscommunication

Connect before you correct.

—APOLLO R. SALLE, BAY MANAGER

ROME, 2013. A Saturday in June. I was enjoying a nice breakfast with friends at an outdoor table in one of my favorite cafes in Budapest's 12th district when I got a call from our station manager in Bucharest informing me that the flight this morning to Rome-Ciampino (CIA) would be diverting to Rome-Fiumicino (FCO), as the pilots couldn't extend the landing gear and a crash landing was expected. At FCO, the airline located better repair facilities, which means, in case the aircraft needed repair. Why do these things always happen on Friday afternoons or Saturdays … seriously!

The landing gear system of a modern aircraft is well protected against malfunctions, and even if you have lost all power on board, you can extend the gears by gravity only. The jargon for that is "free fall," and it is unlikely to fail. But as the flight on this beautiful June day proved, unlikely does not mean it can't happen. In our case, the left-hand door actuator got stuck halfway, and when the pilots activated the free-fall extension, the wheel came resting on the half-open door. At that point, you can't reverse it anymore. So, they had to land on one main gear. Tense moments.

Due to fabulous airmanship, the pilots landed on the right-hand main landing gear and the nose gear enabling the plane to roll on two instead of three legs all the way down the runway until they came to a smooth stop and settled the right wing down on the engine in the grass. Later, it would tilt backward and rest on its tail. Amazingly nobody was hurt, and the aircraft wasn't damaged too badly. The repair at FCO, where we were using Alitalia's hangars as guests, took almost three months. But it could have been so much worse. This was the closest I had ever been to a maintenance-related air disaster.

Our responsibility in this was releasing the aircraft before replacing the problematic actuators. The landing gear door actuators of a certain part number had been found to be unreliable with exactly this behavior and we were required to change them under an EASA Airworthiness Directive (AD)—though, technically, time to comply still existed.

A320 after crash landing at FCO. The aircraft landed on one main gear and came to a rest on the left-hand engine. Passengers disembarked via escape slides. After the evacuation the aircraft tilted on its tail.

Let me explain. All ADs have certain compliance deadlines. The more urgent the problem, the shorter the time to comply. In this case, the landing gear door actuator could seize halfway due to a design or production problem. And although this was obviously an urgent problem, the AD wasn't due yet for compliance. That's no contradiction. Deadlines are set after elaborate risk assessments and, by and large, the aviation system with their regulators at the top, mainly the FAA and the EASA, have an excellent track record for balancing safety with production goals, what means cost. Today, no other mode of transportation is as safe as flying.

So, the limits in the official systems, which track thousands of maintenance deadlines, were not yet reached. Otherwise, the affected aircraft would not have been allowed to fly before the required corrective action: replacement of the actuator.

Technically we were in compliance without the replacement, however, the customer had advised us to check all their aircraft under our care as to whether this particular landing gear door actuator was installed, and if so, to replace it before the next flight. This informal request didn't make it to the team in Romania before the fateful flight took off on that Saturday morning in June.

Today's ease of multiple communication channels became an error trap. If we don't use the official operations system to control a new deadline, we have so many channels which promise immediate and universal dissemination of very urgent instructions: email, text, messenger apps, and company platforms. In the good old days, when communication was expensive and tedious, the SITA telex system was used for all urgent matters, eliminating clutter in the system. Less was somehow more. In our case, the email got lost or wasn't read fast enough. As simple as that.

MISCOMMUNICATION LEADS TO MISHAPS

Miscommunication is at the heart of many, if not most, maintenance mishaps. A good friend of mine in Budapest used to say, "It's all about communication," and he meant "all." So often we are scratching our heads trying to figure out how we have yet another case of miscommunication. Crew resource management, which is concerned with the soft skills of individuals and teams in safety-critical environments, lists "communication" as its first component: communications, situational awareness, problem-solving, decision-making, and teamwork.

It is more than a singular component though. As Professor Helmreich suggests, communication must be present in all other aspects of teamwork, and it can be the "behavioral indicator" of decision-making, problem-solving, and resource and workload management. All these aspects of teamwork influence the way the team communicates with each other, and the way they communicate, in turn, influences the effectiveness of decision-making, problem-solving, and resource and workload management. The effectiveness of a leader's decision depends on the way it is communicated with the team: will members of the team feel entitled to expand or even challenge the decision?

Every single detail in aviation relies on communication. It is more than a contributing factor in mishaps. It is central. The late Professor Barry Turner, author of *Man-Made Disasters*, in the 1970s, summarized it as: "Disaster equals energy plus misinformation." The industrial accidents he investigated all showed some mishandling of information. All the information needed was somewhere available in the organization, but it didn't reach the right team (or was not taken seriously). "Energy" refers to the destructive potential of a plant out of control or an aircraft moving. Sometimes, it just takes a bit of misinformation to release this destructive energy.

We see miscommunication in many of the familiar events, like maintenance limit overruns, for example. A mechanic temporarily repairs a hole with speed-tape and releases the aircraft per Minimum Equipment List, for fifty flight cycles or so before final repair, and then forgets to open a second control for a visual inspection of the tape before every flight. Or towing accidents which I discussed earlier: the information about an impending impact is most often available to the wing walker closest to it, but sometimes they can't get the warning to the tow truck driver fast enough. There's accidental deployment of escape slides. Sometimes one team installs the safety pins on the reservoirs and another team is performing maintenance on the escape slide systems. It may happen that the installation of the safety pins did not happen, e.g., because of a problem with removing the reservoir cover, but the mechanics failed to alert the subsequent team. Now the incoming team might just assume that the safety pins are installed as per routine sequence despite the warning note in the AMM.

"Assuming" is the keyword here. We assume so much in our day-to-day communication. When we order a meal in a restaurant abroad, we become aware of how many things can go wrong, because we just assumed the practices that we know would be universal. We may end up with too little or too much, the wrong sequence, or the most expensive bottle of wine the restaurant offers. In many of our usual endeavors, we take no questions as agreement without making sure that the other party really understands us and that we are all on the same page.

Sometimes the consequences are not as harmless as getting your complete order up to the dessert all at the same time. One of our A380 customers suffered an air turn-back as a result of miscommunication between the flight control (FLC) team and the landing gear (LDG) team. Normally it's the FLC team who fully controls the whole flap system. It just happened here that LDG requested the retraction of the inboard flaps to access the sliding panels in their area obstructed by the flaps. As the FLC team was too busy at that time, they cleared the LDG team to retract the affected flaps themselves.

The LDG team disconnected torque tube assembly no. 18 as part of the process. This escaped the attention of the FLC team. When the LDG team reconnected the torque tube, they only hand-tightened the tube and left the final torquing to a later step. Unfortunately, they did not properly communicate this step to FLC. Once the aircraft was back in operation, the connection gave way after a couple of cycles and the flaps could no longer be fully retracted after takeoff. Imagine 150 tons of fuel being dumped and all the passengers having to stay in a hotel as a result. That was the very costly and embarrassing result of a missed turnover.

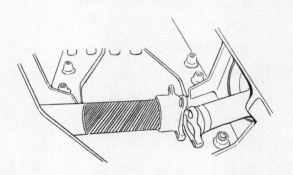

Disconnected torque tube no. 18 on A380 flaps system. The tube separated in flight, as the connection was not finally torqued.

We've reviewed these types of error traps earlier. A temporary installation not properly communicated—we know what happens when the safety stay is not removed before lowering an aircraft from jacks. A final torquing missed. Lack of a proper turnover in the job card or in the shift logbook. All avoidable errors that resulted in costly mistakes. An additional visual cue, like a "remove before flight" streamer, a tag, or some tape could have drawn attention to the unusual state of the aircraft.

Standard procedure should be referring to the official aircraft documentation: the job card. There are robust processes to make sure that job card entries are not overlooked, while the shift logbook gets attention from one shift to the other, it is less suited if closing an item is several days or even weeks away. Additional measures, like the warning tape, are good but may be considered outdated or obsolete by subsequent shifts. In addition to the case of the thrust reverser turnover (the clecos) from above, another case comes to my mind, where the official documentation took precedence.

It was early in my career as an avionic engineer when an obscure fault in one of the Display Management Computers (DMC) of the A320 could not be found. It refused to show on the ground, but it reliably re-appeared in flight. We decided that our best chance would be to have it tested in the full-flight simulator which uses exactly the same computers. Now, to get it installed in the simulator, it had to be fully certified with a serviceability tag or Form 1 (or Form 8130-3 in the US). Without the tag or form, the installation team would likely return it to the shop without installing and testing it. On the other hand, we didn't want it to end up in the flight line, so we wrapped it in warning tape. Something like "only for simulator installation," to no avail. Some colleagues in the logistics chain needed a DMC, saw the Form 1, and just removed the non-standard tape. So, it flew again. In our next try, we hand-carried it to the simulator in Bremen and spent some (fun) hours there, finally determining the scenario under which the fault occurred.

As much as I like the unofficial warning signs, they also must be used with care as they may create a new path of error. We must make efforts to use or adjust the official process as our main tool.

TAKEAWAY NO.16: Use official means, for example, job cards for non-routine situations, to communicate important information for the next shifts. In addition, back it up (but don't replace it) with tape and streamers.

Redundancy Can Minimize Miscommunication

One final thought on redundant directives and warnings. At an Airbus intercultural training for French and German employees, it was established that the Germans tended to over-explain things and that the French found this duplication a bit offensive. For example, on a street within the Airbus facilities in Hamburg there stood a sign stating "Dangerous Road" and beneath it was a second sign stating "Reduce Speed." The French may view this as an egregious over-explanation, but I would argue that in communication, a certain amount of redundancy is good. Don't be shy about reiterating instructions and requirements as needed.

I am not alone in this thinking. Jocko Willink, a best-selling author who coined the term "extreme ownership," found out late in his leadership studies and by chance that people understand much less than leaders assume they do. He shares in his book that he was taken aback by how little of his commander's intent (the bigger picture) is internalized by his staff, despite the extraordinary value he placed on sharing the bigger picture to empower people and subjectively felt he had already over-explained it.

I have found that many leaders are likewise surprised when they learn how "political" their staff views their relationship with them. When many leaders think they have a trustful working relationship with their troops, which may be just one step short of friendship, those troops never forget who has the power. They will always scan every word you say for hidden meaning, and they will always behave in a way that supports their interest at the receiving end of the power gradient, not least of which is keeping their job.

In communication with others, always make the effort to assure that you understand the other party and that they understand you. Redundancy is better than the risk of miscommunication.

TAKEAWAY NO. 17: Don't assume that people understand what you mean or that you understand what they mean.

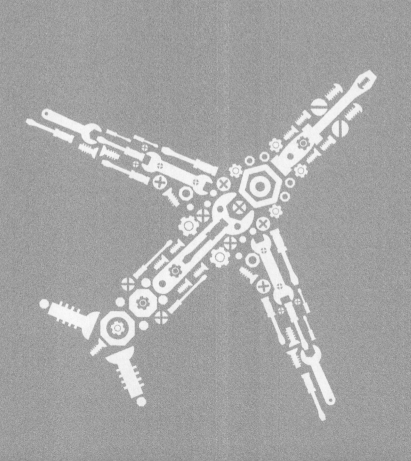

CHAPTER 11

Ticking Boxes Without Thinking

READ & SIGN, JUST SIGN,
OR NOT EVEN THAT?

There is a difference between rush and
urgent. Rush is thoughtless work whereas,
urgent is work done with caution.

—FERNANDO M. COJA, TEAM LEADER

MANILA, 2020. At the start of the pandemic, we were hurrying to get all learning-related activities online and mobile, so that they could be accessed remotely. That included training, qualification, and authorization management. It also included good old read & signs: a staple in aircraft maintenance to disseminate information about company routines, manual changes, and error traps. They were part of almost all corrective actions after quality escapes—reminding people to strictly follow procedures. A classical defense against error traps. In the early days, these written procedures were tracked by lists in which people had to sign off by hand acknowledging the content of each of these often one-pagers, sometimes a-hundred-pagers, which were printed out and put in folders at strategic locations.

As electronic handling of these documents became commonplace, the signatures were replaced by electronic means. In our system, pre-pandemic, people were seeing a backlog in their intranet account of these read & sign documents. To avoid a backlog, each document needed to be opened and an examination question answered. Often, the question may have been as difficult as, "Have you read and understood the document?" Our team was not great about keeping up, and when we needed to change to a modern Learning Management System (LMS), we found that our backlog had swollen to over 20,000 items and the top-notcher was more than 1,000 read & signs late! Nowadays, we (and the LMS) are indeed following up, and we pose an average of five real examination questions per document. A defense, but a clumsy one. By its very nature, even the best read & sign system will not entice people to read it in a way to remember it. They will tick the box.

But, first, we had to get there. We needed a state-of-the-art LMS. All kinds of trainings and subsequent authorizations needed to be done online and in a highly automated fashion. Easy. An LMS can be bought off-the-shelf, and aviation cannot be so special that we can't map our processes into a virtual academy of some sort, right? Well, it wasn't easy at all, but to the credit of our training manager and her team, we were able to launch an online, automated system in July 2020. I'll spare you the painful details.

Professionalism is about striving to get better every day. The sobering reality though is that people don't really enjoy training (except perhaps a high-excitement engine ground-run full-motion simulator training in a foreign country) despite our best efforts and genuine enthusiasm. This may be something particular to aircraft maintenance, but I'm betting that the same disenchantment with training is not unique to us and occurs in hospitals, nuclear power plants, and other industries just as much.

The read & signs contain valuable information about recent incidents, threats, and changes in manuals, which should be easier to digest in a moment of study rather than in the heat of action. They also include reminders, often artfully presented to make consumption easy. Still, many people don't provide the attention needed to this important information. Instead, they just want to get it over with and tick the box.

JUST TICKING THE BOX HELPS NO ONE

The inattention to the read & signs is just the tip of the iceberg. Building enthusiasm for training is a challenge as many, if not most, will tell you that they prefer to work "out there" rather than sit in a classroom. This is unfortunate, as training is such a great opportunity to hone your knowledge without the time pressure of operations. The companies in our industry put enormous amounts of effort and money into these trainings only to have the trainees simply go through the motions so they can say they completed it.

"Ticking the box" means that people show up (even that's not a given) and do what is required to get on the attendance list or pass the exam, which, often enough, is hard to fail anyway. They are not focused on the training. In fairness, they may be concerned about their project in the hangar and its critical challenges, which they can't fix today from the classroom. Maybe they lead a team, and they worry that their deputy may struggle, and they won't be there to assist. It may also be that the proliferation of data privacy and consumer protection waivers and other fine prints we are urged to sign in our daily lives has contributed to a desensitization that conditions us to scroll down to find the "Accept" button.

This conditioning extends beyond the read & signs to virtually all "indirect" activities which are not the hands-on work. It can be observed in:

- Checklists (e.g., docking, housekeeping, earthquake): Check-lists are a great means to avoid omissions, discussed earlier, so that easily overlooked items are readily available, but sometimes more is less here and there is a temptation to tick through the items in bulk.

- Workshops, focus groups, learning teams: All kinds of improvement groups stand and fall with the participation of the front-line employees, but here again, the most qualified of them are often overextended and the real operational work may weigh heavy on their minds.

- Log sheets (e.g., tool control, shelf life, temperature control, vehicle/equipment usage): Log sheets are used everywhere and are too often mindlessly checked off. Have you ever observed in a public restroom, a janitor coming in, signing the log sheet without looking at or doing anything, and then leaving?

- Multiple signatories: Several people signing on the same document without clear distinction of responsibility. This is something what Snook referred to as the "fallacy of social redundancy."

- Pre-briefings, debriefings: As useful as short preparation meetings are, before going out to perform a task together, including an even more important debriefing afterward, these instruments are also in danger of fatiguing employees resulting in them just ticking a box.

- Evaluation sheets: Indispensable feedback instruments after trainings and seminars and all kinds of guided activities, but also in danger of being mindlessly ticked off.

- KPI tracking and performance dialogues: A standard in many operations areas and credited with much higher transparency and honest improvement probing but runs the risk of becoming stale without quite a bit of effort to keep a meaningful exchange going.

- And, finally, audits and investigations: The main tool of the SMS to close gaps and avoid "practical drift." The more standardized they become, the less real attention is raised by the findings and the quicker people want to just move through it.

This long list demonstrates the causes of and the *potential* for mindlessly rushing through our required indirect tasks, but I don't mean to suggest that we are all sleeping through all of our peripheral activities. It is, however, the possible flipside of high situational awareness with the core task at the controls, not least to be able to avoid all the error traps. We must consciously switch from autopilot to manual control in high-risk situations *and in all other situations* including classroom training, briefing huddles, and quality workshops, even though the safety of the aircraft is not immediately at stake. After all, it is these indirect activities that enable us to have the skills and knowledge to successfully switch to manual control in high-risk situations.

Let's, once again, look at our available tools and how they relate to indirect work:

- *Competence:* Know and understand the tendency in our industry to focus all attention and energy to "direct work" and the challenge of all side activities, "indirect work" if you will, to be considered inconsequential and to be done on autopilot.

- *Awareness:* Be in the moment. Nowhere is it as important as in training. If your mind wanders you will not learn. Take all side activities at work as seriously as your core work. If there are non-sensical activities, try to change them.

- *Compliance:* Know "pseudo compliance" when you see it and don't allow yourself to fall into it. Mindlessly ticking boxes doesn't help anybody and can create unnecessary errors.

- *Teamwork:* Help the juniors to see it your way. Help them see the hidden opportunities in training and improvement activities but also the rationale behind housekeeping.

Nothing has changed since our school days:

TAKEAWAY NO. 18: You are not learning for the school or for the teachers, you are learning for your life.

Now that we have addressed the very real challenge of boosting and maintaining interest in trainings and other indirect tasks, let's talk about what organizations can do and what you can do in your capacity as a leader to overcome that challenge. All training, read & signs, and the indirect activities listed above have a better chance of engaging the participants, if we keep them fresh, interactive, and interesting. Easier said than done, of course, but all those activities must be updated and improved regularly. It is the author's, the trainer's, or the facilitator's creativity that will make the difference and there are many great personalities out there who can make classroom activities exciting.

There are a few textbook remedies, and while no one remedy will work in all situations, keep them in your training toolbox and draw on them as appropriate:

- Involvement of the users

- Gamification

- Decluttering

So, what's the next best thing if the textbook remedies aren't useful for your situation and you are short of star designers and star performers to deliver those unloved bits of indirect work, those grueling repetitive, overlapping, redundant sessions which make people dim away? Anything?

No silver bullet, but I am offering to consider this: People, and no generation more than the young one, want to have choices. We are currently trialing a flexible benefit concept in a newly built part of the company where people have choices. They select only those benefits they see real value in for themselves and forego others. This way they are optimizing a certain budget. Without proof and without even much data so far, I would maintain that in a flexible benefit system the employee gets 120 percent of the perceived value for 80 percent of the cost to the company. A clear win-win.

Could that work to better sell trainings as opportunities? Isn't it an age-old sales technique to not ask your client for a yes or no decision but rather a choice between A and B? Consider a mandatory half-day procedures and documentation training (procs & docs). Might engagement be increased if there was opportunity to choose between training options? Option A would be the standard training while option B would be offered as a full-day masterclass for super users who are expected to coach their peers during work? My point is, even if most or all people choose "A," they may feel (a little bit) better about it, simply because they had a choice.

There are countless indirect activities in our industry, which are often not considered "real work," with read & signs the poster child, but encompass many kinds of trainings, housekeeping, and improvement activities. All those are in danger of not being given real attention and simply being ticked off. As users, we must strengthen our awareness of this tendency and monitor ourselves and others to stay in the moment. As leaders, we should look for ways to reinvent the content and delivery constantly. Integrating some sort of choice for the users could help to increase the impact.

CHAPTER 12

Bad News and Second Chances

In their normal way of trying to assign blame, the press asked, "Who makes these solid rocket motors, and where are they built?" NASA responded by telling them that Morton Thiokol supplied the SRMs from its plant in Utah. [...] The next morning a friend at the Utah plant told me that someone had spray-painted the words "Thiokol—Murderers" on the side of one of the overpasses on the main road to the plant.

**—ALLAN J. MCDONALD, *TRUTH, LIES,
AND O-RINGS: INSIDE THE SPACE
SHUTTLE CHALLENGER DISASTER***

MANILA, 2007. We were growing and improving, but I was not happy. We simply produced too many quality escapes, technical incidents, and inexcusable delays, which is something particular to airframe maintenance. Our branch of MRO has a long and deep tradition of apportioning blame, often accompanied by hefty penalties. Fierce customer "on-site" teams watch us 24/7. Nevertheless, we were about to take the next step and sign up big world-renowned brands as our aircraft overhaul customers. I was just coming back from Australia and went straight from the terminal to the hangar. My boss thought I was micromanaging the operations. But we had a great leadership team at that time, and we were committed to conquer the world. In the hangar, I was immediately hunting for the reasons why we had lost two days while I was away to win more customers. That's not micromanaging. I was looking for a face of the slack and that face happened to be a poor fellow who was just coming back from what looked like an extended break. I confronted him—even physically. Terrible. He was totally innocent, and he had just moved his break to later due to work-related reasons. That was me, but also it wasn't. That was somebody who punished the messenger or any other unlucky fellow who happened to be at the wrong time in the wrong place. Something had to change.

And seeing it wasn't the bad news that wanted to change; I had to change.

A decade later, when we worked with world-class aviation safety scientists like Sidney Dekker and RJ de Boer, they directed our attention to how we take in bad news and how that helps or hinders learning. The default reaction is to get mad with the apparently stupid perpetrator. How can somebody drive the flaps into the docking platforms, mix up bar and psi, or forget to remove the aft stay at dejacking?

Another safety scientist, Todd Conklin, offers the following anecdote:

> "They called him "Captain," and when he came into the training room, he took the place by storm. … Captain was a safety leader at his company. He was also an excellent worker. … I was so surprised when he got angry. … He did everything but stand up and call me an idiot. … From the back of the room came this statement, 'You have to admit

that some people are just stupid. … If you tell a worker, not to stick his finger into a hole every day and one day you leave, and he sticks his finger in that same hole you warned him about and he cuts his finger off. That guy is just an idiot, he's stupid.' Captain was convinced that he was right on this, and I was wrong."

Pre-Accident Investigations: Better Questions—An Applied Approach to Operational Learning by Todd Conklin, p. xiii, CRC Press 2016

Conklin makes a controversial point here. He refuses to accept that the stupid guy was indeed stupid. Captain's opinion could well be considered the mainstream. How stupid can one be? But Conklin explains that people in safety-critical environments, with time pressure and goal conflicts around them, make mistakes, not choices.

The difference is powerful.

Rational choice is deeply ingrained in Western civilization. Decisions follow logical, secular, unbiased reasoning based on scientific assumptions to achieve an optimal output benefit. Different options are compared, and the most promising course of action is selected. This understanding is also reflected by the FOR-DEC decision-making model for flight crews: Facts are established based on which Options are identified. These Options are assessed for Risks and based on this first stage, a Decision is made, which is then Executed and finally Checked for effectiveness.

This sounds intuitive, but that such a model exists and is used in training, points to the fact that this might not be the default mode of decision-making in practice. We looked at Kahneman's "thinking fast and slow" earlier. Rational choice as ideal is "slow thinking," and that's why it must be trained. Our default mode is "fast thinking":

no risk assessment and only one option based on perceptions instead of facts. For most of history, the survival of our species depended on split-second decisions to fight or flight. Today we don't even need real-time pressure to invoke this archaic dynamic, and it is indeed our default way to get through the day, and so we call it "naturalistic" decision-making. But it is error-prone, to be sure. Otherwise, we wouldn't train the pilots in FOR-DEC.

In naturalistic decision-making, intuition plays a big part. We process the facts, or merely the perceptions, based on past experience, patterns which we quickly recognize, and past successes and failures. Rationality may be sacrificed for speed. Here again, we can also fall off the other side of the horse. Consider a site selection decision. Is it better to invest in Vietnam or in Thailand? The big spreadsheets of weighted factors look extremely rational, but the selection, scores, and weights of the factors are highly subjective, although the math itself doesn't lie. Many management decisions are like that, too complex for FOR-DEC. Even more so for operators of high-risk systems and indeed pilots: Neil Armstrong saved Gemini-8 from spiraling out of control by firing the re-entry thrusters. Was it intuition or an extremely fast FOR-DEC process? Second pilot David Scott later said, it was his "lucky day, that Armstrong knew the systems so well."

You can look at every decision from both angles. But it seems that, at the end of the day, some decisions are too fast, too intuitive, that's why we train FOR-DEC. And some decisions are too slow, when we struggle to foresee what comes next in a complex system or when we must rank our values which are all equally dear to us. And even if the situation is saved by the right mixture of cool reasoning and intuition, beware of complexity—even Armstrong, one of the best test pilots of his time, could only do so much and needed some luck that in such a multi-failure scenario the rest of the system behaved in a benign way. And that has only become worse since software has taken control of large chunks of our systems, and multiple failures can produce software states which have never been seen before, let alone support a benign recovery.

Now that I've finished my brief detour to the manned space program and the heroes of my childhood, let's return to the concept of choices versus mistakes. Rational choice is our revered ideal, but in real life, we often don't have the time and all the information required, and even then, we don't make the effort because we have other important things to get done. When do mistakes happen? Isn't it, that in this flow of "naturalistic decision-making" we forget things, misjudge situations and consequences, or try to do two things at the same time? Most of the time it works well, but not always.

Errors can also happen after a thoughtful deliberation. People rob banks after a lot of thinking. If the values or the goals we pursue are not good, a risk assessment won't put it right. Garbage-in-garbage-out as they say in software. But this is not what happens when a ramp vehicle crashes into an aircraft. There is no planning, no risk assessment—no rational choice. It's a mistake.

TAKEAWAY NO. 19: People make mistakes, not choices.

WOW: A RAMP COLLISION IS ONLY A MISTAKE?

Don't get me wrong. It's not that mistakes are good. When I was visiting one of our main customers for our Puerto Rico operations at their headquarters in New York, we discussed staff identification numbers and how to deal with a mechanic losing their stamp (a mistake). They said they give that staff a new number with a Dash-One (-1) suffix (a second chance). If they lose their ID or their stamp again, they will get a Dash-Two (-2) added to their original number (one more chance). There are no Dash-Threes. I think that captures it beautifully.

Giving people a second chance is as much part of Western civilization as rational choice. Granted, sometimes life doesn't give you a second chance, but more often than not, we decide whether to give a person their Dash-One badge. Professor Reason remarked that often the biggest mistakes are involving the best people. Of course, big mistakes don't automatically make great people, but many high-caliber aviation professionals have mind-boggling disasters on their record books—they just can't do that too often and neither can you.

How do we get that message across? I'd like to think that this doesn't require much deliberation most of the time. People with a clear, professional attitude suffer the most from their mistakes themselves, and often lack the words to explain it. They don't need somebody to amplify it. If a player misses a decisive penalty kick, none of the teammates would explain to them how it should have been done. They would console the player and would re-assure them that "failure is not fatal" in Churchill's words or in some other fashion.

How we learn from our mistakes shapes and shows our attitude and should lead to not repeating it and furthermore, promoting the right lessons firsthand. The reality is, we can't make mistakes too often and, if we do, we may need to find a job outside aviation. That might sound harsh but, on the other hand, to imagine second chances offers a lot of leeway. It doesn't answer whether there could be one single violation of work practices grave enough to rule out even a second chance. It is meant for the majority of incidents that we have to deal with, not the extreme cases. Those incidents, which bring about the bad news to a possibly hotheaded production leader.

Let us look at the moment of truth. Somebody screwed up. The aircraft crashed into ground equipment or vice versa. Or we find out we have used the wrong material after the twentieth project. Or the pilots are already on board, and we can't find a panel that was removed at the beginning of the layover. The bad news comes in all shapes and colors.

What to do and what to avoid?

1. *Solve the problem and its consequences first.* In our anger we want answers. Who did what? How could it happen? That's all secondary. Concentrate on minimizing the damage and repairing it, and that typically also calms down the upset customer.

2. *Wait for the investigation to be closed.* Don't speculate. Trust the process. Often it is much different from how it first appears. Give people the benefit of the doubt until the investigation is really done. This is often not easy with upset customers, but this is industry standard. No airline would start to speculate about the reasons for a crash before official information from the investigation is available, and too often we have seen "facts" reported on day one, turn out to be false later.

3. *Make it an organizational problem and an organizational solution.* Consider that the involved people could have been other people as well, producing a mishap under similar circumstances. Make the system stronger and don't look narrowly at this particular scenario but do look at the obvious weaknesses as symptoms of deeper problems which could also manifest at a different place and in a different form if not tackled at the root. Customers appreciate it if the incident

is not handled as an isolated "one-off" or exclusively blamed on the operators. Customers expect the organization to go beyond and to rectify more than meets the eye.

While this may sound perfectly straightforward, all three steps are in danger of drifting into the blame game. The crucial step is waiting for the complete investigation. Too often, blatant human error looks much differently when the full picture is known. The process requires patience on all sides including the customer, so that expectations for the corrective actions can evolve and be met. The correct action process requires serious effort and innovative thinking to make the system stronger.

Like "aviate, navigate, communicate," this sequence, which we might phrase "mitigate, investigate, innovate," might also help to keep priorities straight and our own emotions in check on the day of the incident. There will be enough time for emotions once the investigation is complete. Disciplinary action may be in the cards by then, but only then.

TAKEAWAY NO. 20: Suppress your anger. It's not too late to get angry when you have the full picture.

After more than twenty years as a production leader, I dare to say that bad news is best dealt with by giving the people involved the benefit of the doubt, counseling them, and directing your energy on:

TAKEAWAY NO. 21: Mitigate, investigate, innovate.

At the same time, let's show our young generation of mechanics how closely we operate at the edge of catastrophe every day. They must be aware of the error traps, understand how they emerge, and why even in a very safe system, we can't eliminate them altogether. We can help them with some standard tools, but they must find out for themselves what works best for them.

And the same outside work. Our lives are full of error traps. And we make the same mistakes just as scores of people before us did. In most accident investigations, we see more than one warning sign, more than one "red flag," if it had just been seen as such. We fail to see them in everyday life just the same. Instead, we deviate from good, accepted practice without thinking (casual non-compliance) or we decide it is not applicable to us (conscious non-compliance), and so we build up all kinds of bad habits—eating the wrong stuff, letting our relationships deteriorate, not thinking long term—aggravated by wrong tactical decisions when it counts most and inspired by ego, envy, and greed. But at the heart of it, all of it is predictable and preventable, just like the incorrect closing of the V2500 fan cowl latches.

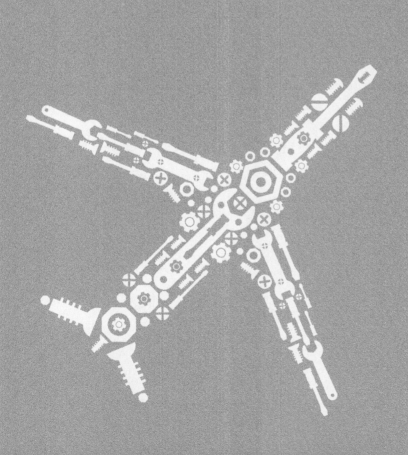

CONCLUSION

*A safety culture is a culture that allows
the boss to hear bad news.*

**—SIDNEY DEKKER, THE FIELD GUIDE TO
UNDERSTANDING HUMAN ERROR**

IN THE AVIATION MAINTENANCE INDUSTRY, *books* don't seem to be particularly popular on the shop floor or in the offices, in junior or in senior circles. Maybe this is a sign of our YouTube times, or maybe it is because people in our industry are haunted from day one to "work by the books"—a connotation almost as bad as "work to rule." Being an eternal optimist, I consider this an opportunity for this book to stand out. Maybe this is the only book you bother to read in a long time. And if you have made it here, this is the point to admit that there are more books out there which have laid out possible improvements in aviation safety much better than I could. The problem was that too few people read them. I didn't want to burden the text with references that distort the flow. But as we summarize the argument of *Error Traps* and try to answer: "So what?," now is the time to give credit to the sources and encourage the esteemed reader to follow up on one or the other topic with one of those excellent texts.

It all starts with human factors. I consider error traps an under-valued aspect of human factors, but it hardly makes sense without the main body of knowledge. The Australian aviation authority, CASA, has published a comprehensive overview of human factors for engineers, which is free, at least in Australia.[1] From there, we move further into human error: a good general textbook is Reason and Hobbs' *Managing Maintenance Error*,[2] which applies to aviation maintenance the earlier work on *Organizational Accidents*.[3, 4] And, finally, Dekker's *Field Guide to Understanding 'Human Error'* and *Just Culture*,[5, 6] which challenges the mainstream thinking mainly rep-resented by Reason. Further recommendations are provided below.

Why try to make an already safe system—or ultra-safe system, as some would call it—even safer? Is it worth our efforts or can only a technological leap forward get us to the next level?[7] Isn't flying with modern airlines already so irrationally safe, compared to other modes of transportation, that we should direct our energy elsewhere?[8] I disagree, because the (remaining) mishaps we are suffering from follow such distinct patterns and appear so highly preventable[9] that I can't look the other way. What is more, those *error traps* are waiting to impose misery on real people, perhaps unsuspecting young mechanics under my watch. James Reason[2] wants us to find and eliminate our error traps, but let's face it: that won't be possible, not in aircraft maintenance and not in life. We build defenses, but sometimes too few, sometimes too many. Sometimes we build clumsy defenses and sometimes sophisticated ones, which the users don't understand and unknowingly undermine.

Perhaps that is the crux: Our mechanics have some understand-ing of human factors; they know that time pressure or fatigue can make errors more likely. But they often don't understand the defenses, or they don't take them for what they are. They look at "following procedures" as a tool for standardization, performance management, or satisfying legal obligations. But they often don't recognize Standard Operating Procedures (SOPs) as protection against error traps. Following procedures or not exposes our mechanics enormously, not only for the potential to weaken the defenses, but also for the possibility to be blamed for incidents and accidents. This could be fair or unfair, and different organizations have better or worse *just cultures*. The exposure remains, however, especially if people deviate from guidance deliberately without a good reason.[5, 6, 10, 11]

I used the example of the V2500 FCD closing as the *mother of all error traps*. Despite several attempts to erect new defenses, such as painting the latches or requiring a techlog entry, incidents kept occurring roughly once per year with huge financial and reputational consequences. In 2018, a comprehensive solution was found to redesign the latches and to require a key with an attached flag to be held in the latch until it would be closed again. But even this was not a guarantee for avoiding the error trap, as it doesn't seem to be very intuitive to mechanics and at times, one key was used for both sides, despite the warnings in the manual. Finally, although this seems to be such a prominent case with its frequent high-profile damages, most people have never heard of it.

There are error traps like the FCD and defenses all around us, which are ignored or circumvented. It is part of our industry and part of our life. The operators, but also the managers, designers, planners in aviation, hospitals, and power plants should keep in mind that we must learn from mishaps and make the systems more forgiving in the face of human performance variability,[12, 13] and acknowledge that error traps will continue to be present, despite our best efforts.

Error traps and imperfect defenses abound. This is really the main message, which I believe should complement our human factors toolkit, for mechanics and all of us as operators of some sorts. We looked at error traps in design, tools, processes, and communication. It often boils down to complacency, and, yes, professional sloppiness. And, of course, we don't stop there. We try to make the system stronger and more robust at every turn. If users don't understand and circumvent defenses, we try to improve them.

But we can never fully rely on the system to protect us from error traps, and what kind of work would that be anyway?[14] What to do? Probably everybody must find out what works best for them. I have suggested the four tools highlighted by Tony Kern: *Competence*, *awareness*, *compliance*, and *teamwork*. But this is just one way to look at it. If you are with me, that all these error traps can't be eliminated, we must adjust our behavior, one way or another.[15]

Lots of effort goes into making the "hazard controls" more independent from human performance or the lack of it. And as all this will still not be enough, from time to time, we must repair the damage. We should strive for a humane way to do it[16] and we should take full ownership.[17]

So, how do high-performing teams learn to avoid mistakes in aircraft maintenance? This is a simple and a complex challenge at the same time. It is simple but not as simple as "we have to tell people to be more careful," which is still in many leaders' minds. If we can just overcome this, we have already achieved something. But it is still simple enough to summarize that high-performing teams learn to avoid mistakes by constantly expanding their understanding of the world (competence), which enables them to maximize their mindfulness when it counts most (awareness), and at the same time use standards as much as reasonably possible (compliance), while their members support and back up each other in all three aspects (teamwork). And all this is complex at the same time, as our dark side works against it and can obscure and corrupt all four dimensions. Our competence breeds hubris, arrogance, and complacency. Our own awareness or task orientation may wear out our patience and empathy with less focused colleagues. Compliance comes under constant pressure of optimization for a more efficient way to do things, and, finally, teamwork is always vulnerable to bad role models, power dynamics, and the influence of the wider system. Life is full of error traps. It takes leadership to avoid them, and nobody said that would be easy.

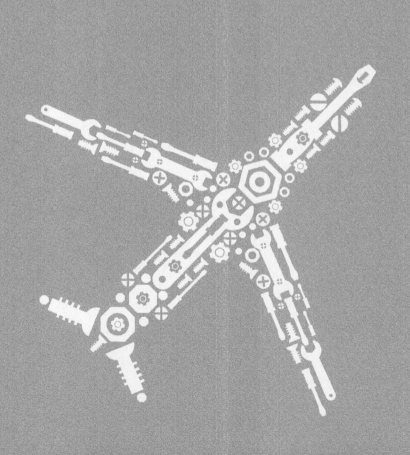

ACKNOWLEDGMENTS

I would like to sincerely thank my editor, Beth Cooper, for her professional, flexible, and inspired counsel, and the team at Forbes Books including Nate Best, Megan Elger, Jacob Hollifield, and many others, who believed in the idea and enthusiastically helped to bring this book to life. Furthermore, I would like to thank Leila Jamolangue, who made the illustrations and dealt patiently with frequent revisions. My sincere appreciation goes further to leading safety scientists and practioners Dr. Sidney Dekker, Dr. R.J. de Boer, and Dr. Tony Kern, who have all worked with us on site to make aircraft maintenance even safer. Most of their insights I have embraced wholeheartedly. Some bits remain controversial, but that's the spice in the debate. I am also endlessly grateful for the support in my own organization, Lufthansa Technik, and here especially to Tara Mariano, Marc Bambei, and Dr. Juergen Dollmayer, who initiated the Pamana in-house training program, and Apple Pangan and Joey Virola, who curated the MyLegacy series. And special thanks to the many high-profile industry experts I could talk to in the course of this project. Finally, I would like to thank Eunice Gan, Philipe Janke, and Dea Bonita in my own office, not least for taking care of all the logistics, but also for our daily discussions on the topic.

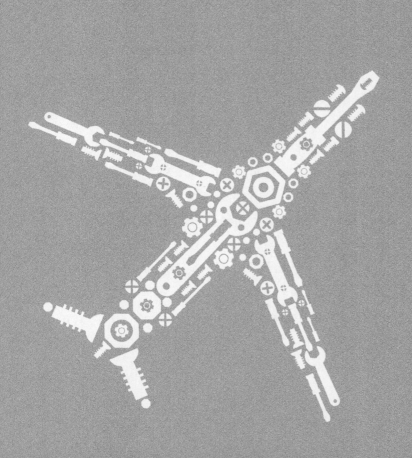

GLOSSARY

AD	Airworthiness Directive
AMM	Aircraft Maintenance Manual—One of several manuals released by the aircraft manufacturer that guides maintenance works. The AMM describes all general tasks on the aircraft itself, while more specialized work on components, structures, or electrical installations, or using NDT is described in dedicated manuals for these tasks.
AOG	Aircraft on Ground
AUTOMATION SURPRISE	When an automated system behaves in ways that the operators do not expect.
ATA	Air Transport Association
CB	Circuit Breaker
C-CHECK	The C-check is part of the maintenance program every aircraft has. Every operator (the airline) must get approval for their maintenance program by their aviation authority. The C-check is a hangar visit between 4 days and 4 weeks, and takes place every 1 to 3 years depending on type, age, findings, and modification status of the aircraft.
CML	Consumable Material List
CRM	Crew Resource Management—A concept dealing with the "soft skills" of *teamwork* and communication in the cockpit, later extended to the whole crew, maintenance, and other work teams outside of aviation. The poster child of CRM is reigning in an overbearing captain without flirting with mutiny.

D-CHECK	See also C-check. The D-check is the biggest hangar visit of an aircraft, every 6 to 8 C-checks, or, typically every 12 years. It is sometimes combined with a landing gear change and can take up to 3 months. The term "D-check" is not used anymore in official documents though, as today even the heaviest checks are referred to as C-checks.
DMC	Display Management Computers
EASA	European Aviation Safety Agency
EGR	Engine Ground Run
EPC	Error-Producing Conditions (attributed to Tony Kern). EPCs with acronyms: HR/LF High-Risk/Low Frequency, LSNR Low Signal to Noise Ratio, NOD normalization to deviance.
ETOPS	Extended Twin Operations—Aircraft with two instead of four engines are given additional maintenance tasks and requirements to further reduce the probability of a dual-engine failure, when they fly extended time out of reach for an alternate airport to land, typically over water such as trans-Atlantic and trans-Pacific.
FAA	Federal Aviation Administration
FCD	Fan Cowl Door—A pair of cowlings or movable panels protecting the installations mounted on the engine case in the area of the fan or low-pressure compressor. The FCDs are closed by a number of latches (depending on the engine type) in the lowest position close to the ground. Improper closing might lead to separation of the FCDs in flight.
FLC	Flight Control
FMU	Fuel Metering Unit
FOD	Foreign Objects Debris
FOR-DEC	Decision-making model for flight crews: Facts are established based on which Options are identified. These Options are assessed for Risks and based on this first stage, a Decision is made, which is then Executed and finally Checked for effectiveness.

GSE	Ground Support Equipment
HAZAT	Hazardous Attitudes (attributed to Tony Kern)
HP	High Pressure
HTP	Horizontal Tail Plane
ICAO	International Civil Aviation Organization
IDG	Integrated Drive Generator
ISO	International Standards Organization
JOB CARD	A tailored description of a maintenance task prepared from original documentation by a technical support office. It is also referred to as a Task Card or a Work Card.
LDG	Landing Gear
LMS	Learning Management System
LP	Low Pressure
MCAS	Maneuvering Characteristics Augmentation System—A piece of the Flight Control Computer software of the B737MAX aircraft, which was intended to correct an excessive nose-up behavior of the aircraft in certain flight scenarios due to the bigger and more outward engines compared to the B737NG. This correction was implemented as a command to the horizontal stabilizer, initially not revealed to the pilots.
MCD	Magnetic Chip Detector—A device to collect metal chips in engine oil that point to contamination of the fluid. MCDs must be checked and changed frequently in line maintenance, which opens and closes the oil system and, thus, introduces the risk of oil loss if not done correctly. Oil loss in flight could lead to engine shut down.
MFTF	Movable Flap Track Fairings
MRO	Maintenance, Repair, and Overhaul

NDT	Non-Destructive Testing—An array of inspection methods that allow inspectors to evaluate and collect data about a material, system, or component without permanently altering it.
NTO	No Technical Objection
NTSB	The National Transportation Safety Board
OEM	Original Equipment Manufacturer
PDA	Part Departing Aircraft
PME	Precision Measurement Equipment
QUALITY ESCAPE	Any product or service containing a deviation/defect that is released from point of origin, whether caught before it reached the end external customer or not.
SMS	Safety Management System
SOP	Standard Operating Procedure
SRM	Structural Repair Manual—One of the specialized maintenance manuals dealing with the repair of the aircraft's metal or composite structure.
TA	Technical Adaptation
TAT	Turn-Around Time—The time required to "turn around" an aircraft, which means repair or service it, either in line maintenance at the terminal or in base maintenance during a hangar check. These TATs determine the performance of the maintenance organization. Short and reliable TATs define performing organizations.
THS	Trimmable Horizontal Stabilizer
TPS	Temporary Protection System
VPC	Violation-Producing Conditions (attributed to Tony Kern)

REFERENCES AND FURTHER READING

[1] CASA. "Safety behaviours: human factors for engineers kit." CASA Online Store. 2013. Accessed April 21, 2023. https://shop.casa.gov. au/products/safety-behaviours-human-factors-for-engineers.

[2] Reason, James, and Hobbs, Alan. *Managing Maintenance Error: A Practical Guide*. London: CRC Press, 2017. DOI: 10.1201/9781315249926.

[3] Reason, James. *Managing the Risks of Organizational Accidents*. London: Routledge, 1997.

[4] Reason, James. *Organizational Accidents Revisited*. London: CRC Press, 2016. DOI: 10.4324/9781315562841.

[5] Dekker, Sidney. *The Field Guide to Understanding 'Human Error,'* 3rd ed. London: CRC Press, 2017. DOI: 10.1201/9781317031833.

[6] Dekker, Sidney. *Just Culture: Restoring Trust and Accountability in Your Organization, Third Edition*. London: CRC Press, 2016. DOI: 10.1201/9781315590813.

[7] Amalberti, R. "The Paradoxes of Almost Totally Safe Transportation Systems." *Safety Science* 37, no. 2 (2001): 109–126. DOI: 10.1016/ S0925-7535(00)00045-X.

[8]   Perrow, Charles. *Normal Accidents: Living with High Risk Technologies*. New York: Basic Books, 1984.

[9]   Kern, Tony. *Blue Threat: Why to Err Is Inhuman*. Pygmy Books, 2009.

[10] Cribb, Alan, Jane K. O'Hara, and Justin Waring. "Improving Responses to Safety Incidents: We Need to Talk about Justice." *BMJ Quality & Safety* 31, no. 4 (2022): 327–330.

[11] Department of Defense. "Human factors analysis and classification system (DOD HFACS) version 7.0." Department of Defense. 2017. [Online]. https://www.dcms.uscg.mil/Portals/10/CG-1/cg113/docs/pdf/DoD_HFACS7.0.pdf?ver=2017-02-23-152408-007.

[12] Hollnagel, Erik, David D. Woods, and Nancy Leveson, editors. *Resilience Engineering: Concepts and Precepts*. London: CRC Press, 2017. DOI: 10.1201/9781315605685.

[13] Hollnagel, Erik. *Safety-I and Safety-II: The Past and Future of Safety Management*. London: CRC Press, 2014.

[14] Dekker, Sidney. *Foundations of Safety Science: A Century of Understanding Accidents and Disasters*. London: Routledge, 2019.

[15] Goldsmith, Marshall, and Mark Reiter. *Triggers: Sparking Positive Change and Making It Last*, Main Edition. London: Profile Books, 2016.

[16] Dekker, Sidney, Amanda Oates, and Joseph Rafferty. *Restorative Just Culture in Practice: Implementation and Evaluation*, 1st ed. New York: Productivity Press, 2022.

[17] Willink, Jocko, and Leif Babin. *Extreme Ownership: How U.S. Navy SEALs Lead and Win*. New York: St. Martin's Press, 2017.